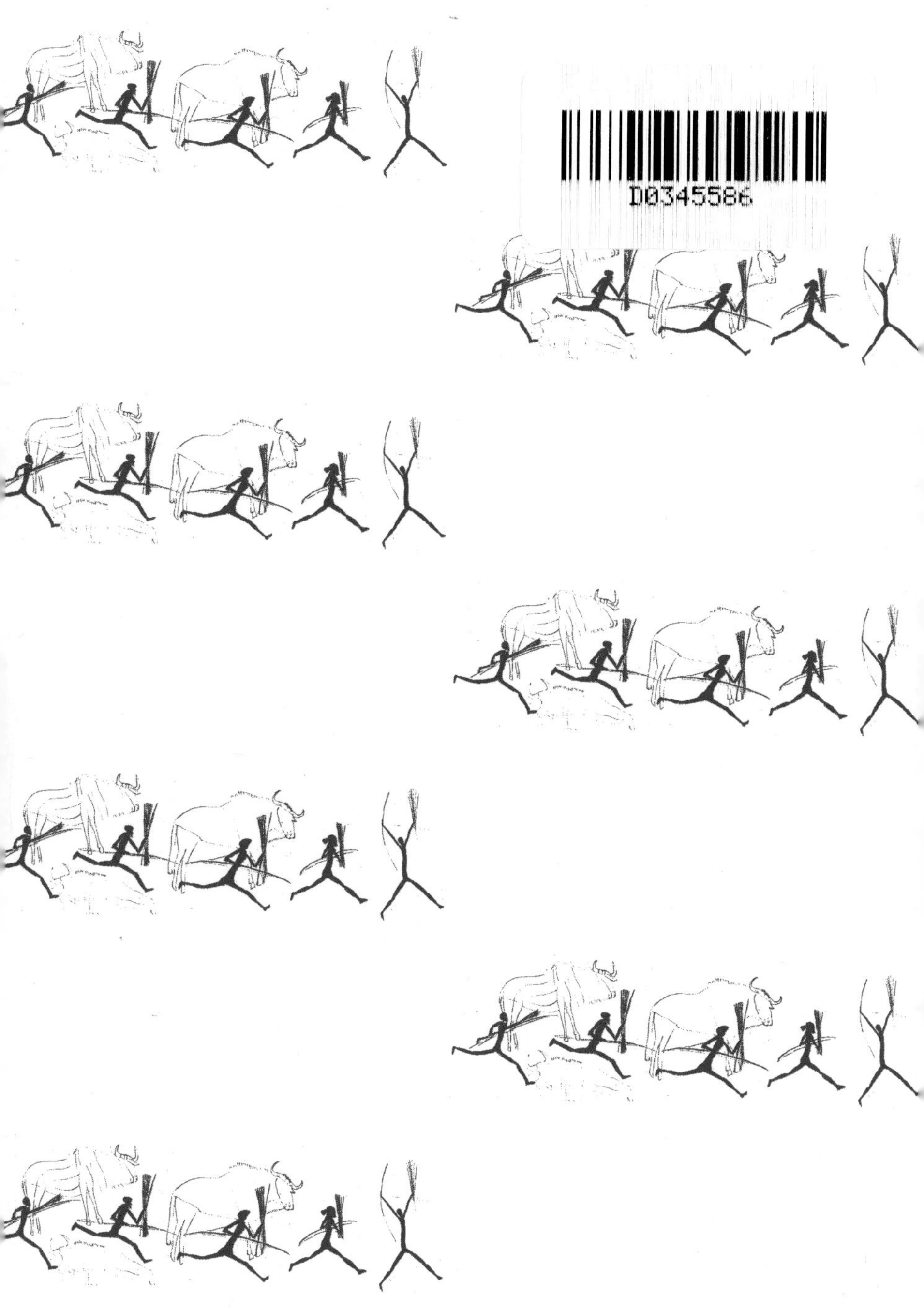
D0345586

Racing the Antelope

Also by Bernd Heinrich

The Trees in My Forest

A Year in the Maine Woods

Ravens in Winter

One Man's Owl

Bumblebee Economics

Mind of the Raven

Racing the Antelope

What Animals Can Teach Us About Running and Life

Bernd Heinrich

Cliff Street Books
An Imprint of HarperCollins*Publishers*

HarperCollins books may be purchased for educational, business, or sales promotional use. For information please write: Special Markets Department, HarperCollins Publishers, Inc., 10 East 53rd Street, New York, NY 10022.

FIRST EDITION

Designed by William Ruoto

Library of Congress Cataloging-in-Publication Data
Heinrich, Bernd, 1940–
Racing the antelope: what animals can teach us about running and life / Bernd Heinrich.—1st ed.
p. cm.
Includes bibliographical references (p. 269).
ISBN 0-06-019921-0
1. Running. 2. Physiology, Comparative. 3. Human evolution. I. Title.
QP310.R85 H45 2001
612'.044—dc21 00-047287

01 02 03 04 05 ❖/RRD 10 9 8 7 6 5 4 3 2 1

To Erica, Stuart, Eliot, and Lena,
and to their enduring mothers.

Contents

Acknowledgments

I'm grateful to Bob Colby and Edmund Styrna, my high school and college coaches, whose selfless devotion to ideals of excellence inspired me on and off the track. The late Dick Cook—teacher, scientist, and friend—gave me the encouragement I needed to begin. I attribute the inspiration to keep going to innumerable runners, of whom my college teammates—especially the late Mike Kimball, and Jerry Ellis, Bruce Wentworth, Kirk Hanson, Horace Horton, and Timothy Carter—deserve special mention. We ran together for four important years. I thank Bill Gayton for his devotion and sense of humor as a race director who made a certain twenty-four-hour run memorable, although I here focus on a different engagement. I owe sincere thanks and gratitude to David Blaikie for his help in retrieving data and giving encouragement, and to Kim Layfield for her invaluably fast and efficient typing.

All runners are deeply grateful to those who can help them spot flaws to improve pace, strength, and economy of stride. Writing is like running. It requires seemingly innumerable repetition to improve pace, strength, and economy of words until it all looks easy. Thank you so very much, Sandy Dijkstra, Diane Reverand, Alice Calaprice, Bill Patrick, and especially my wife, Rachel Smolker, who diligently cleared the way to rescue me from many a stumble.

Prologue

The human experience is populated with dreams and aspirations. For me, the animal totem for these dreams is the antelope, swift, strong, and elusive. Most of us chase after "antelopes," and sometimes we catch them. Often we don't. But why do we bother to try? I think it is because without dream-"antelopes" to chase we become what a lapdog is to a wolf. And we are inherently more like wolves than lapdogs, because the communal chase is part of our biological makeup.

For me, the glimpse of a new "antelope" on the horizon

came in early May 1981. I had seen a fresh sign, and I had to give chase. I had just run my first ultramarathon, a 50-kilometer race, a short race that barely qualifies as an ultramarathon. But in the final half mile I had passed the then-current U.S. National 100-kilometer record holder, which made me wonder if, just possibly, I had the potential to race well at long distances. The North American 100-kilometer championships were to be held on October 4 that year in Chicago. Although at that moment I could just barely have run another step further, I began to dream about the potential of racing the 100 km, twice as far as I had ever run before.

The problem was: how to prepare to run that far? As a zoologist by profession, it seemed only natural for me to look to other "endurance athlete" species to see why and how it's done, and for tips on how to train. However, I did not write this book as a training manual, nor did I write it to highlight my running exploits, which are puny relative to those of others. I wrote to show what is involved in running an ultramarathon race, and to pull together the race experience with the insights I gained from my studies of animals. My intent is to amalgamate the race experience with human biology to explore what makes us different from other animals, and in what ways we are the same. In the process, I discovered some possibly new perspectives on human evolution.

ONE

Wind-in-the-Face Warm-Up

I love running cross-country. . . . You come up a hill and see two deer going, "What the hell is he doing?" On a track I feel like a hamster.

—ROBIN WILLIAMS, *film star*

These days, my daily run is almost always in the mode of a wind-down after a long day of sedentary activity. I come home feeling a little restless, and eager to smell fresh air and as I change into running shorts and a light pair of running shoes, I start to feel new. I feel transformed and free, like a caterpillar molting into a butterfly. Seconds after tying my laces, I can trot down the driveway.

It is overcast this afternoon (September 21, 1999) and there is a fine, misty drizzle that feels fresh on my face. The still air amplifies the sound of water dripping on maple leaves. The leaves are still bright green, but they will transform into a kaleidoscope of yellow, orange, red, salmon, and purple in another week or two. The goldenrods along the dirt road are just starting to fade, and several species of wild asters are flowering instead. I note the splashes of their lavender, purple, and blue flowers. There are usually bumblebees on these flowers, but today these cold-hardy bees remain torpid in their underground nests deep in the woods.

Watching a large orange and black monarch butterfly feeding at an aster, I wonder how much sugar it is getting from the nectar to fuel on this stop on its migration from Canada to Mexico. The butterflies, like human ultramarathoners (those who race 50 or more miles), need regular refueling stations. While it was warm and sunny during the last couple of weeks, I've daily seen the monarchs floating by on lazy, soaring wing beats. These individuals are at least the third generation of those that left central Mexico last spring to come north to breed. All of them are now journeying to their communal wintering area in the cool mountains near Mexico City from where their ancestors had come. There they conserve their energy reserves through the winter by literally putting themselves in refrigeration that slows their metabolic fires. What incredibly long journeys these delicate creatures make just to avoid lethal freezing, while keeping themselves at a low-enough temperature to conserve their energy supplies during months of fasting! Monarch butterflies are long-distance travelers. It is in their makeup. It is their way of coping.

I turn left at the bottom of the driveway, just across from the beaver bog. It is quiet there today. In April I'd heard the cacophony of the snipes whinnying and the red-winged blackbirds yodeling, and all were gone already two months ago. Dragonflies emerged from their larvae in the cold water, seeking warmth. But today, all the dragonflies, their muscles cold, are grounded. Mist collects in droplets on their wings as they perch limply on the cattail foliage. I glance across the bog to the beaver lodge in the pond where the Canadian geese nested. One never knows, there could be a moose, a great blue heron, otters. . . . No moose and no geese today. Anyday now, any hour, the geese's haunting cries as they glide through the sky will signal the birds' excitement as they, too, head south, arranged into long *V*s. Like human runners following one another's wind shadow, they take advantage of reduced air resistance to save energy.

Almost everything we know about ourselves has been built on knowledge learned from other organisms: Gregor Mendel's peas, George Beadle and Edward Tatum's bread mold, Barbara McClintock's corn, and Thomas Hunt Morgan's fruit flies have taught us the basics of inheritance. Mice, rats, dogs, and monkeys have been the subject of studies that provide us with an endless knowledge of practically all our physiological functions. From studies of rats and mice we learned how to fight viruses, battle bacteria, and guard against debilitating diseases. Without insights gleaned from other animals in their natural environment in the field, knowledge of our behavior, our psychology, and our origins would be superficial and rudimentary. As Koyukon elder Grandpa William told anthropologist Robert Nelson (in *The Island Within,* Vintage Books, 1991), "Every animal knows more than you do." So I too believe that animals can teach us much about running. They've been doing it for many millions of years before there were recognizable humans.

We can find animals who are far superior to us in practicing what we preach in terms of industry, fidelity, loyalty, bravery, monogamy, patience, and tolerance, but looking to other animals in order to justify our own moral codes is dangerous. Their example can be used just as easily to justify hate, violence, torture, cannibalism, infanticide, deception, rape, murder, and even war and genocide. They can show us how we became what we are, but not what we should try to become. We can learn from them about running the way *we* want to run.

Given the grand diversity of animals on this planet, we are hardly more unique or even special than most others. We are the product of a vast evolutionary grandeur being created under the same interplay of innumerable constraints and possibilities. Only through them can we see ourselves objectively through an otherwise vapid haze of wishful thinking and unbridled assumptions.

Past the pond, in the five-foot-thick, half-dead sugar maple, a hairy woodpecker hammers on the thick, dry branches, ignoring the jogger. Nearby, a flock of robins swiftly scatters from the young maple trees overgrown with wild grapes. The birds are doing last-day fattening up for their migration, feasting on the berries that conveniently ripen at this time. A grouse feeding on the grapes the robins have knocked to the ground explodes in a loud whir of wings. Its powerful, swift flight startles me. If flushed repeatedly, the grouse would tire and become unable to fly. Like most migrants, the robins can fly nonstop for hundreds, possibly thousands, of miles. They can show us the many specializations required for endurance. Grouse have what it takes for explosive power. I presume they have fast-twitch fibers, like champion human sprinters.

Less than a quarter mile farther down the road, I come to another beaver habitation, this one only a year and a half old. The new dam has flooded the woods, and this summer the flooded trees are dying. The beavers are felling large poplars along the edge of the pond to make underwater food caches of twigs to sustain them through the coming winter. As I jog by the pond, wood ducks quickly paddle through the new surface covering of green duckweed, to hide in the flooded winterberry shrubs. The berries are already red, signaling their ripeness to migrant birds. The wood ducks had nested nearby in a cavity carved out by a pileated woodpecker, and within hours after hatching out of the egg, the ducklings jumped like Ping-Pong balls straight up to the entrance and out. They are natural-born high jumpers. The family of wood duck youngsters grew up in this pond last spring. In May, they were tiny downy chicks; now the whole family looks like adults.

Every few steps, I notice caterpillar feeding damage on the leaves of the trees' branches over the road. Earlier I had seen caterpillar droppings on the smooth road surface, and I'd look up to find big green cecropia and other moth larvae that have now pupated to pass the winter in torpor. Few caterpillars are visible now, but soon the nests of robins and vireos that were hidden all summer will be exposed as the leaves start to fall from the trees. In the beaver bog, the dying maples already have brilliant yellow and orange leaves. These dying trees have peaked sooner than the others and invite admiring glances.

I'm getting warmed up now. My stride is lengthening and becoming looser. I'm feeling better, thinking clearer, and remembering things I had long forgotten. Just past the beaver pond I pass the steep overhanging bank along the road where I'd found a veery nest in the spring. I saw the bird's dark eyes as she hunched down, watching me pass. The nest is deserted now—but it conjures up images of bright blue eggs, pink, naked young sporting a light white fuzz, and clumsy spotted youngsters who later hopped in front of me on the road.

A few more steps—the beech tree. There are no capsule fragments on the road to indicate that squirrels had been feasting, so there is likely no beechnut crop this year. On the other side of the road are the old-growth hardwoods, where I've often seen a barred owl.

Another few steps to the turn in the road by the apple tree, where some deer jumped over the road, and where I once saw two young beavers doddering along.

Next comes a level stretch of about half a mile. I speed up slightly, mentally trying to visualize every portion of my stride. I'm surprised how conscious visualization of my stride seems to affect it. Keep the movement smooth—keep track of exactly what the right side does as opposed to the left at the same moment. Thinking about what I'm doing, I feel it, then feed the information from mind into body and back again. Back and forth. OK—I've got it! Like most of our knowledge, such action is usually unconscious.

At the farm pond, I veer off the road and jump over a fence to check for frogs and see how the pond is refilling with the recent rains. I'd heard tree frogs here last month, and I'd seen mink and green frogs. Like some hunter exploring the veldt, I expect to see many treasures on my runs.

Jogging on again, I look up toward Camels Hump mountain, then descend a long slope down to the Huntington River. I'm feeling good, empowered by what's around the bend, by memories of past runs, and sometimes also by the lure of a race in the future.

Two

Ancient Runners and Us

Yet that man is happy and poets sing of him who conquers with hand and swift foot and strength.

—PINDAR, *Greek poet, c. 500 B.C.*

. . . The essential thing in life is not so much conquering as fighting well.

—BARON PIERRE DE COUBERTIN, *on reviving the Olympic Games in 1896*

We are all natural-born runners, although many of us forget this fact. I will never forget when I first ran barefoot as a child on the warm sand of a lonely wooded road in Germany, where I smelled the pines, heard wood pigeons coo, and saw bright green tiger beetles running or flying ahead of me. I will never, *never* forget running on asphalt pavement on October 4, 1981, more than thirty years later. On that day I raced a 100-kilometer distance in Chicago with 261 other men and women. Each of them was in one way or another, like me, chasing a dream antelope. When I began to think about what running is all about for us humans, and why I raced, I was surprised at the vividness of my distant memories, and at my new revelations. There were many worlds between the small boy running barefoot on the sand and the forty-one-year-old biologist wearing Nikes on the Chicago pavement. But now these memories were intertwined in my mind with the larger scheme of human existence that relates to our kinship with animals and goes back to the dawn of humankind. Those thoughts gave new meaning to this race.

Movement is almost synonymous with life. With elongating stems and twirling tendrils, plants race one another toward light. Similarly, the seeds of many plants compete to be first on the right piece of ground. Some may travel hundreds of miles by ingenious and diverse mechanisms: being carried by wind or water, or being ferried by berry-eating birds or fur-bearing mammals.

Animals move primarily on their own power: they harness chemical energy by means of muscles. But like plants, we humans have recently harnessed the wind, water, and other animals to carry us. And increasingly, our species, unlike any other, is tapping the energy from coal, oil, and the atom for locomotion.

Throughout the hundreds of millions of years of animal evolution, there has been selective pressure on some species to be able to travel farther and quicker, and to do it more economically and under ever more adverse conditions than either their competitors or their predators. Both predators and prey have to move faster or die. An anonymous runner captured the notion in this now-famous aphorism: "Every morning in Africa, an antelope wakes up. It knows it must outrun the fastest lion, or it will be killed. Every morning in Africa, a lion wakes up. It knows it must run faster than the fastest antelope, or it will starve. It doesn't matter whether you're a lion or an antelope—when the sun comes up, you'd better be running." Of course, these animals don't need to know—they must only be fast.

With the help of our infinite imagination and the technologies it has produced, we now travel faster, more economically, and well beyond the range of our muscle power. But for millions of years, our ultimate form of locomotion was running. We are, deep down, still runners, whether or not we declare it by our actions. And our minds, as much as our lungs and muscles, are a vital force that empowers our running. Whenever one of us jogs down a road or when we line up to race in a marathon, we are not only celebrating life in general and our individual aliveness but we are also exercising our fantasies while acknowledging reality. We are secure in the knowledge that there is no magic. Which is not to say the world is only of simple logic, because although it may be simple in its design, it is awesomely complex in its details.

I've run at varying distances and intensities almost all of my life, probably because the primal unadorned simplicity of running appeals to me. Various games incorporate running, but only running itself touches the pure and basic essence of the tension between speed and endurance, stripped bare of our everyday world of technology, beliefs, and hype. Nothing—*nothing* in the world, in terms of sheer performance—compares in my mind to the thrill of seeing a Lee Evans rounding the curve on the way to a 400-meter finish, or the electricity of a Peter Snell, a Cathy Freeman, a Billy Mills, or a Joan Benoit Samuelson on the homestretch to an Olympic victory. Why? Because it is pure and powerful.

The Complete Book of Running by the late James F. Fixx concludes with the following lines:

> My suspicion is that the effects of running are not extraordinary at all, but quite ordinary. It is the other states, all other feelings, that are peculiar, for they are an abnegation of the way you and I are intended to feel. As runners, I think we reach directly back along the endless chain of history. We experience what we would have felt had we lived ten thousand years ago, eating fruits, nuts and vegetables, and keeping our hearts and lungs and muscles fit by constant movement. We are reasserting as modern man seldom does, our kinship with ancient man, and even with the wild beasts that preceded him.

Several years ago in Matopos (now Matobo) National Park, Zimbabwe, I had a rare opportunity to experience this feeling of kinship with ancient runners that Fixx alludes to. I was on a research trip to study how body temperature affects the running and fighting ability of scarab dung beetles. On the rolling hills, their rock outcrops covered by short grass, I saw and smelled white and yellow flowering acacia trees that were abuzz with bees, wasps, and colorful cetoniid beetles. Giraffes were peaceably grazing on the flat-topped acacias. Baboons and impalas, each in their respective groups or bands, roamed in the miombe bush. Tens of thousands of wildebeest and zebra can, at the right time of year, still be seen in such a landscape, thundering by during their massive migration. Elephants and rhinoceros lumber like prehistoric giants over the land, ever on the move. Serendipitously, I looked under a rather inauspicious and small rock overhang and was taken aback by what I saw.

The cave painting

Painted onto the wall under the overhang was a succession of small, sticklike human figures in clear running stride. All were clutching delicate bows, quivers, and arrows. These hunters were running in one direction, from left to right across the rock face. In itself, this two- or three-thousand-year-old pictograph was not particularly extraordinary. But then I noticed something more, and it sent my mind reeling. It was the figure farthest to the right, the one leading the progression. It had its hands thrown up in the air in the universal runners' gesture of triumph at the end of a race. This involuntary gesture is reflexive for most runners who have fought hard, who have breathed the heat and smelled the fire, and then felt the exhilaration of triumph over adversity. The image of the Bushmen remains for me an iconic reminder that the roots of our running, our competitiveness, and our striving for excellence go back very far and very deep.

Looking at that African rock painting made me feel that I was witness to a kindred spirit, a man who had long ago vanished yet whom I understood as if we'd talked just a moment earlier. I was not only in the same environment and of the same mind as this unknown Bushman running hunter, I was also in the place that most likely produced our common ancestors. The artist had been here hundreds of generations before me, but that was only the blink of an eye compared to the eons that have elapsed since a bipedal intermediate between our apelike and our recognizably human ancestors left the safety of the forest for the savanna some 4 million years ago, to start running. There is nothing quite so gentle, deep, and irrational as our running—and nothing quite so savage, and so wild.

THREE

Start of the Race

I see you stand like greyhounds in the slips, straining upon the start.
—SHAKESPEARE, *Henry V*

For me, the Bushman painting embodies the connection between running, hunting, and humanity's striving toward excellence for its own sake. All other animals are much more strictly utilitarian. They lack that artistic drive which is detached from ulterior motives and rewards. Looking at the Bushman painting, I thought of the late Steve Prefontaine, from Coos Bay, Oregon, former University of Oregon runner who was one of the greatest and gutsiest all-time middle-distance racers ever. Pre put it this way: "A race is like a work of art that people can look at and be affected by in as many ways as they're capable of understanding." Yes, the key to appreciation is in the understanding.

As I was growing up, my gods were runners like Herb Elliott, Jim Ryun, and the now-anonymous men on competing teams who could outrun me. These weren't just people. Some of them seemed to defy natural laws. My appreciation came from the understanding that what they did was extraordinary, and not readily understood. All I knew was: it was not magic. I wanted to know what they ate and breathed and how they lived, what made them so different from other humans and so much like some of the animals I admired.

Seeing a great performance, whether by a human or another animal, still inspires me to no end. I'm moved by others' dreams and by their devotion and courage in the pursuit of excellence. I get choked up when I see a kid, or anyone else, fighting hopeless odds—someone who goes out there to run the lonely roads with a dream in the heart, a gleam in the eye, and a goal in mind. I admire those who have the courage to step up to the line of a great race to run their heart out for a dream. I empathize with a heart touched by fire during this Dream Time of youth, when as runners we were still undefeated in spirit, felt invincible, and thought the world was pure.

Many of the people close to me in rural Maine did not appear to strive for great things. I saw them carrying their black lunch pails, with a thermos and a sandwich, each morning on their way to the dreary bowels of the clanging woolen mill. In the evening they came back, milked the cow, and went to bed. After some years of the same endlessly repeated routine, they died, usually in the same hospital where they were born.

I wanted to do something different. However, that is a difficult thing if you see no opportunity. On the other hand, it is hard not to try when you think you can do something when you have a chance at success, even though it is often hazardous to strike out on one's own. That seldom goes unpunished. Any mark of difference may become a target. Even my own father, to whom I owe so much, had taught me this harsh lesson.

His eyesight was failing and he could not carry on much longer with his work as an entomologist. He wanted me to continue in his footsteps in ichneumonid taxonomy to perpetuate his dream. But I had my own dreams, in a different world. He was a fine field naturalist, but his interpretive skills were not based on modern science in which I was being trained. I recall the day when, near the end of my undergraduate studies, I was home on a short vacation, sitting near him at his desk in the old farmhouse where he spent hours each day peering through his microscope, "preparing" specimens. That preparation involved meticulously pinning the wings and legs of every insect on a piece of cork, using thin long insect pins to hold the appendages in place until they dried. He spent hours each day to prepare two or three of his specimens. Every single one in his collection of thousands had every single one of its six legs, two antennae, and four wings in the same precise orientation as every other one. After the insect dried and the pins were removed, he put the specimen into one of the neat rows, each with a label with printed dates, location, and so on in almost microscopic script.

Papa was a veteran of two world wars, who had sacrificed his formal education by enlisting for military service at age seventeen because of ideals—some prince of an ally to his country had been assassinated, and it was his duty to defend his country and its sacred commitments. During my school break, I asked his advice on whether I should also enlist, to fight in Vietnam. I don't recall the precise words—except the last ones—but the conversation went something like this: "America is an experiment," he said, and after a long pause continued, "where the driving force is individuals chasing money. I would not risk my bones for a society guided by this principle."

I felt this to be a put-down because I had worked hard to earn money, to finance my education, and to buy a beat-up car. I loved being an American. "The experiment seems to be working out just fine," I said, thinking of the optimism and the well-being I felt and saw all around.

"But it is not done yet," he continued. "Money brings ease, ease makes softness, and throughout history it has always been the toughest, most self-sacrificing who have survived and conquered." So I went to the army recruiter in Bangor, Maine, to try to enlist in the paratroopers.

"Who should we serve anyway, if not ourselves?" I had asked my father rhetorically, adding that I thought he, too, sought selfish satisfaction through his ichneumon flies, and that "maybe we serve the greater good by serving ourselves."

He made what I thought were inappropriate analogies to social insect societies with which I strongly disagreed. Then he was silent for a moment, put down his forceps, looked me in the eye, and soberly declared: "If you don't think like me, then you are not my son," and silently resumed his work. His view of our having to think alike to be liked seemed extreme then, but perhaps it is not. It is usually only better camouflaged.

I became a scientist in part because I sought some measure of certainty in a world where values were all too often defined on the basis of stature, individual bias, unproven assumptions, wishful thinking, dogma, and sentimentality. However, even in science there are often no hard, universally accepted standards that apply outside one's strictly defined discipline. One's greatest theory is another's so-so generality. Another's greatest experimental empirical triumph is another's trivia, if it does not fit into some preconceived "useful" framework. This is not out of malice but from a motivation for excellence, because we as human beings are limited but don't know our limits.

Ultimately, running appealed to me because its quality cannot be defined in terms of anyone's *use* or place in a hierarchy or plan. Perfection is fairly and objectively defined by numbers. There are strict levels of excellence that anyone who chooses can easily recognize and aspire to, with the ultimate being a record. There are rules to the game and the number that one may achieve—whether time taken to run a certain distance, place in a race, or a record—is never open to judgment. Nor can it be snatched away, falsified, or claimed by anyone else. The test is the race, where credentials mean nothing and performances everything.

I've been running since before I was ten years old. At age forty I suddenly became mindful of Pindar's ode to an Olympic winner—"Brief is the seasons of man's delight"—and also the fact that, according to world expert running physiologist David Costill, "there is no doubt that a distance runner is at his best between the ages of 27–32 years." When I turned forty-one, in the spring of 1981, I took stern cognizance of life's trajectory. I hung a wild new hope on a lifelong dream that filled my gut with fire and my mind with a stubborn faith and optimism. It was still possible. I planned to race, and possibly win, the U.S. National Championship at 100 kilometers, to be held in Chicago that fall.

Go for it now or you'll regret it forever, my mind said back then. This act of will was not my last chance to be alive, but it sure felt like it. Running had been a small part of my life. But every part is important, if it is going to be a part at all.

The Chicago race was billed as a battle royal between two ultrarunning greats, Barney Klecker from Minnesota and Don Paul from San Francisco. Klecker, at age twenty-nine, had just set a phenomenal world record. He had run 50 miles in under five hours—in 4:51.25, to be exact. Paul was close in ability, and also hungry. In my mind, Klecker and Paul were invincible. They were human antelopes—swift, unbeatable runners with muscular thighs, thin lower legs, and deep chests.

What would happen when Klecker and Paul—and the rest of us—raced over a distance of 100 kilometers (62.137 miles)? All of us would confront our personal limits, but since Klecker was the very best the world had so far produced, then the limits of *human* speed and endurance for that distance would be challenged as well.

My new bride, Margaret, had flown with me to Chicago the night before the race, but we didn't go to the prerace clinic where dignitaries and name runners would speak and where Don Paul would predict "an exciting race." Instead, I checked out the starting line, jogged along a section of the racecourse on the sidewalk along Lake Michigan, and then retired to our hotel. I lounged in a bathtub filled with hot water. All summer long we had lived in the Maine woods in a tiny tar-paper shack without electricity and running water, studying insects and a tame great horned owl and preparing for this event. This bath was a treat I could not miss.

When I got up the next morning, I ate as many yeast rolls as I could hold and drank a big cup of coffee from a thermos, and then as the dawn showed on the horizon, we hurried down to the starting line.

Figures were darting in and out of the shadows, stretching and striding to warm up as light rain and gusts of wind blew in from the lake. I walked around in my cotton warm-up suit, shivering and nervous. I couldn't wait to take off. I couldn't wait for the anticipated relief at the end of only a few more hours, after months of unremitting daily training at a pace faster than the race's. In my mind, most of the race had been in the 1,500 miles or so I'd run that summer and early fall, and the many tens of thousands in the decades before. By a rough calculation, I had run a distance of at least four times around the globe. Only 62 more miles to go!

Tension mounted as we lined up in rows behind the bold white chalk line drawn across the black pavement. I wondered how many others there were, in the crowd of 261 starters from all corners of the United States and Canada, who had thought as long about this moment, trained as hard for it, and were as energized by dreams as I was.

As I learned later, there were still other dedicated runners lined up besides Klecker and Paul. There was Park Barner, for one. Park is an ultrarunning legend who had regularly run the remarkable distance of 200 miles per week in training. Dan Helfer of Morton, Illinois, was back. He had run second to Klecker in his record run in the 50-mile last year. Roger Rouiller was also in the lineup, a veteran of sixty-three marathons and the American masters' (in the over-forty age group) record holder for 50 miles. Among the women was Sue Ellen Trapp, the American record holder for 50 miles. As a total newcomer, I had heard only of Klecker and Paul, and they were magnified larger than life in my mind. Unanticipated and totally unbeknown to me at the time, it was a race of the best North American ultrarunners.

Paul and Klecker and most of the others were lined up in front of me. The tips of the front of their running shoes were almost touching the white line. I hung back and tucked myself, almost hiding, into the crowd. I found only one familiar face. I stood next to Ray Krolewicz, who had driven out from his home in Pontiac, North Carolina. I had just met him the previous evening as we both checked out the starting line. I didn't know it then, but Krolewicz was also a veteran, who had already raced in more than sixty ultramarathons. I had raced in only one. Solidly built, he looked as tough and indestructible as a camel. Last year he'd been third in the 100-kilometer race held at this same place.

We continued to mill around, anxiously stretching, doing last-minute adjustments of shoelaces, and glancing at our watches. With minutes to go, we crowded ever closer behind the line, awaiting the start with eager trepidation. Some of us anxiously stripped off our warm-ups. But cooled muscles unload oxygen more slowly from the blood, reducing their capacity for high rates of power output. From a study of tiger beetles and the work I had done in the field in Africa, I knew cold beetles run much slower than hot ones, so I waited.

Finally someone, presumably Dr. Noel Nequin, the race director, stood up with the bullhorn and announced race directions. Only seconds now. We tensed. I stripped off my lower warm-ups and threw them aside. Then came the *bang*. The line surged forward, released like fingers letting go of a tensed bowstring. Klecker, Paul, and many others took off at what to me seemed like a frightening pace. Like a wildebeest on the savanna, I became part of the thundering herd behind them.

As the first miles rolled by, I found Krolewicz beside me, talking a blue streak. I couldn't listen. I became lost in streams of consciousness and in long periods when introspection reached back to near unconsciousness: I achieved a runner's trance. At times I tried to energize myself by attempting to think, to motivate myself with pep talks, and to retrieve the Cat Stevens song I'd memorized to accompany and soothe me on the run. But all I could recall was the beat and a few snatches of words: "I've been running a long time, on this traveling ground . . . eons come and gone . . ." Then the words would stop, and I saw flickering images of what seemed, indeed, like eons come and gone.

FOUR

Back to the Beginning

Every parting gives a foretaste of death, every reunion a hint of resurrection.
—ARTHUR SCHOPENHAUER, *German philosopher*

How does a grown man convince himself to spend money and precious time to run himself half to death along the Chicago lakeshore? I asked myself this question during the race, even as I have probably asked it a hundred times since. I came up only with rationalizations. Now, as I look back, I know that part of the answer is that I simply love to run. Perhaps that love seeped into my bones back in my childhood, when I ran on that sandy wooded road chasing shiny, metallic-green tiger beetles. I owe much to that tranquil sylvan existence, where the forest was my playpen. I must therefore take you back there with me, to revisit my forest.

I remember a visit to that special place, touching down

at daybreak on a flight from Boston, then an hour or so later, traveling on a bus toward Trittau over the flat northern German countryside. I'm returning for the first time to see a world that had for years become (or remained) a dreamlike fairyland in my memory. But Trittau does exist—I see a sign that tells me so, then some shops, linden trees, thatched roofs, bright redbrick houses, and tidy grainfields. Coming down a slight hill, I see something familiar about the curve of the road.

The bus stops. I'm in Trittau. I stand dazed, trying to get my bearings. At the corner, just across the street from the bus stop, is a Chinese restaurant, where once stood the grammar school I went to, with its dreary, bare classrooms and a big gray yard. I remember the boy on the bench next to me who had spilled some ink, who was called up in front of the class, where he had to extend his hand as the teacher whacked it repeatedly with a ruler. The boy didn't make a sound.

The hotel across from where the school used to be is still there. Customers are sitting around tables, ordering beer already at noon. The row of horse chestnut trees along the nearby stonemasonry wall looks familiar, too. I imagine a thin, undersized boy who, from ages five to ten, had lived in a one-room cabin with his family in the adjoining large forest. To me he seems like someone else. With his sister, Marianne, who was one year younger, he daily walked or jogged by those trees, which marked the end of their morning two-mile excursion through the woods to town. Once he'd taken a train ride with his father from here to the nearby city of Hamburg. There was bombed-out rubble as far as the eye could see. Nothing green. He remembered hearing about people who, after the bombing and firestorms, got trapped underground, for years, in a storeroom containing nothing but bags of flour. They buried their dead in that flour to squelch the stink of rotting flesh, as they died one by one. He remembered the frightening sound of airplanes. Ever since, he thought that cities were targets, to be avoided at all costs. He loved the forest. Was that boy really me? If so, I would remember those woods. They would not have changed.

The forest was then called the Hahnheide (literally, Rooster's or Fowl's Heath), though the name would later be changed to Schwarmarner Schweiz and it would be designated a nature reserve. I am eager to revisit my old haunts, but I first walk across the once cobblestoned street, which is paved now, to have lunch first. I sit at an unoccupied table in the courtyard at the hotel, order beer, schnitzel, and fries from the busy waitress. I try to remember what the forest looked like, how it smelled, sounded, and felt.

Where is that tiny cabin hidden deep in the forest where we lived so many years when I'd been a companion to crows and a collector of carabid beetles I had called *Lauf,* or running, beetles? Wisps of memory come back. Like the fragmented notes of a song learned long ago, they flicker and then fade. Each wisp of memory leads on to the next stanza.

I pick up my knapsack and walk the now almost familiar route along the stone wall and the horse chestnut trees. Less than a hundred yards later, I'm at an old redbrick gristmill by the linden trees along the pond. This mill had once been powered by water from the dam with a huge wooden waterwheel. Here we sold the beechnuts we had gathered in the forest, which were later pressed into oil for margarine.

I am surprised to see the old brick mill still standing next to the pond, but in more detail and with enchantment I remember coots and gray green-footed, secretive moorhen that had nested in the dense tangle of phragmites and willows along the edge of the mill pond. Sedge warblers sang here, and I had found magic in the several kinds of bird nests that were revealed in my eager explorations.

My steps speed up as I continue up the slight rise to the Stolzenberg place. This estate had been our destination when I was a four-year-old and our family had fled the much-feared Russian troops advancing from the east. We'd miraculously arrived here unharmed, having experienced a series of the most unlikely and improbable adventures, which for decades thereafter etched and almost defined our identities.

How lucky we had been! We escaped the Russian army in a three-month journey from near Gdansk that included leaving in the night, riding in a horse-drawn sleigh and, before the end, also a wagon pulled by two horses, a truck, a cattle train, a sojourn with an outgunned and surrounded German army panzer tank unit, and escaped in the wreck of a Junker's airplane with only one propeller that almost didn't take off and that got shot at. We made it! Other refugees, who had flooded west earlier, when the going was easier, had arrived before us. Instead of a room at the Stolzenbergs', who were acquaintances of my father, we got temporary shelter in an open shed in a nearby cow pasture. It was spring then, and close to the end of the war. It was from that shelter that Papa and Mamusha explored the forest and found the abandoned one-room hut that would be our home until we came to America, to settle on a decaying farm in Maine.

Back then, I had walked or jogged by the Stolzenberg mansion on the way to and from school. The place had always looked spooky and deserted, covered with ivy and surrounded by old cherry trees growing in the unkempt yard. On the second floor lived a boy slightly older than I, who scavenged pieces of lead by leaning out the top window and chewing it off the drainpipe. We used the lead to shoot birds with the slingshots we'd made from carefully selected tree limb forks and red rubber from a discarded inner tube. Frau von Gordon, who like us was also a refugee from East Prussia, had a room on the ground floor. She smoked cigars made from homegrown tobacco, and she walked in a slouch. Her three sons and her husband had all been killed in the war.

I have no idea who is living in the house now, but I'm still drawn to it. It looks even more dilapidated than I remember, not surprising for the passage of decades. I walk along the brick path under the same old cherry trees, and hesitatingly knock on the large wooden back door. There is no answer. I knock again. Slowly, laboriously, wooden-sounding steps descend stairs. A pause. The door opens a crack. An ancient woman peeks out, looking at me blankly. I tell her in German that I'm one of the five Heinriches (sometimes six) who lived for six years after the war in the one-room hut out in the Hahnheide. She stares, remains silent, and closes the door. Perhaps her life is still tinged with danger.

I walk on along the path by the old railroad tracks, toward the tiny station house where the monstrous black steam-whistling locomotive stopped. Both the track and station house are gone, but a bike path remains. I had come by here twice daily. One day I had made the round trip twice. Papa had sold some wood from pine stumps he had dug out of the ground, and I'd been given cash and told to stop and buy bread at the village bakery on my way back from school. But I came back empty-handed. Forgetting was no excuse. I had to turn around and go back again, in order to learn that if you don't have it in your head, then you must have it in your legs—a good theory, but more running never helped me get less spacey. Just possibly it made me a better runner.

My intuition tells me that it takes no special effort of

running early in life to become a great runner later. There are hardly more impressive runners than Bruce Bickford, who grew up on a farm in central Maine and who did not do much exercise beyond plenty of farm chores until he took up cross-country in his sophomore year of high school, when he immediately became a world-class runner. Similarly, Joan Benoit Samuelson of Freeport, Maine, one of the world's greatest women marathoners in history, did not formally begin to run until high school, at age sixteen. On the other hand, Andrew Sockalexis, a famous runner in the early part of the twentieth century, took up running at age ten, when his father, on the Penobscot Indian Reservation at Old Town, Maine, built him a running track near his house to train on. Perhaps running, unlike weight events, involves relatively little restructuring of the body from what it is designed to do already, given the genetic raw material, proper nutrition, and a few simple instructions. The question is, what is the raw material and what are the environmental triggers, or "instructions"?

I don't remember all my instructions, but I do remember one day as it was getting late when I was on my way home to our hut and I saw a man in the road up ahead of me. I bolted in fright into the woods to make a major detour because I had never seen anyone on "our" road before—it was a path leading to nowhere. I was late because I had been dawdling, gnawing off several flakes of crust from the bread I had bought. I knew this to be a lapse in morals since I was never supposed to eat anything except what I was given at meals. I was not anxious to face the consequences of having given in to temptation. It was then, with bread crust in my mouth, that I had dreamed of paradise, a place where you could eat whatever and whenever you wished. I distinctly remember the place in the road where I was thinking of fried chicken as possibly the defining ingredient of heaven, when the man in the road loomed up ahead. Ultimately, however, I remember this forest as a heaven because of its insects, plants, and animals that all became magnified in my mind, so that I saw them with wonder in their exquisite and beautiful detail.

Now I hurry up the hill past the house where Forester Grützmann used to live. Seeing Marianne and me pass by here twice a day, he used to call us Hansel and Gretel. Forester Grützmann had for many years reared caterpillars in cages in a shed by his house. Papa, whose passion was collecting parasitic ichneumon wasps (commonly called "flies"), had taken the wasps that sometimes emerged from the moth pupae and started a new collection. His old collection, representing his life work, had been left buried in metal boxes somewhere in a secret hiding place in a forest back home in East Prussia (now Poland), from which it was retrieved intact decades later. Papa later sent Forester Grützmann some moss specimens from Maine for his moss collection. The forester had been important to us because he owned a shotgun, and he took Papa out to hunt birds. I usually tagged along, and I'd sometimes act like a retriever to recover fallen birds. Mamusha then skinned the birds and made them into museum mounts. Papa then sold them to the American Museum of Natural History, in New York, and other museums. Although the war was over, Germany was still shunned by the world. No mail and no foreign travel were allowed. An intermediary, a Dutch friend, mailed our catch overseas. During these outings, I saw up close for the first time the marvel of birds, a beauty that was only suggested by seeing them from a distance. With Grützmann we also hunted caterpillars in the summer by laying a sheet under a small tree and then whacking the tree, catching whatever fell out. Surprising creatures always fell out of the trees, and what I learned in those years still enriches my life.

Herr Grützmann, being the forester, had a car. One day he drove it slowly up the sandy road toward our cabin as Marianne and I were coming home from school, and I ran eagerly alongside. It was summer, and I was barefoot as always, feeling the warmth and soft texture of the sand with my toes. I often ran on my way to and from school, although I had to stop frequently to watch ants or sand wasps, or to wait for Marianne to catch up. I always felt like running, and on that day I raced Forester Grützmann's car. When we got to the bend in the road, where the hidden footpath led into the forest to our cabin, I was still with him. He stopped the car, got out, and seemingly surprised to see me, he made some comment about my running.

Memories flood back. Looking down, I almost expect to see bare little feet. But it's not the late 1940s. I'm wearing my dirty Nike Mariah running shoes with dark blue trim that I wore in the 100-kilometer race in Chicago. Each shoe has three razor-cut holes I added for extra ventilation and to reduce weight at the front. Along the side, I wrote in already fading ballpoint pen the finishing times of my best races. These numbers represent important points of my life, and here I suddenly see the past and the present merge and connect.

Movement is the essence of life. I moved then because I wanted to go from one place to another. I did it with my legs. So did other creatures. My favorites, the carabids, or running beetles, ran fast in a beautiful coordinated stride, somehow making their six legs work in precise coordination. Most carabids were nocturnal predators. However, one group of them, the beautifully iridescent green cicindelids, or tiger beetles, were active in the day. They seek sunshine. I saw

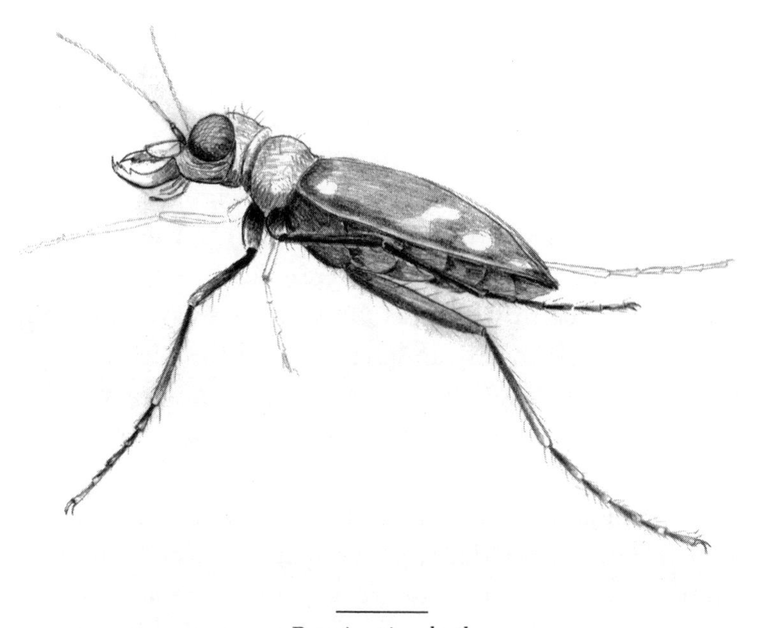

Running tiger beetle

them in the spring on this same sandy patch of our road. There were many of them. When I came close to one, it would run so fast that its thread-thin legs were a blur. When I got closer still, it would take off, flying close over the sand down the road ahead. I'd often speed up, trying to outrun the beetle as it raced on like a bright green jewel. The beetles flew faster than I could run, landing far ahead. I would catch up and begin the chase again, but I could never quite catch one of these beetles on a warm and sunny day. On overcast days they did not generally show themselves, but if one did, it was at a disadvantage. Not heated up by the sun, it was a slow runner, and it could not fly at all. The beetle, which before had easily outdistanced me, was now suddenly within my grasp. After discovering their Achilles' heel, I captured one for my growing beetle collection, among which the running beetles were my favorites.

Leg speed in beetles is, as in humans, dependent on both body build and muscle temperature. African dung beetles, in their large diversity of body build and thermal strategies, provide an excellent demonstration. There are some species that remind one of chunky, muscular weight lifters with their round bodies and thick, short legs. Their leg speed is very slow, but their *strength* is so great that they can tunnel through solid earth with ease. Other species have thin legs, and these gracile forms run fast, provided their leg muscles are at a high enough temperature. Running velocity increases 400 percent as temperature increases from 28°C (82°F) to 35°C (95°F) when the fastest dung beetles achieve velocities of about 25 centimeters per second. Tiger beetles run five times faster than this, even at the same body temperatures, probably because they have much longer and spindlier legs than even the fastest dung beetles. Tiger beetles stay in sunshine to maintain a high body temperature by basking, and if they are hot enough they may fly instead of run, thereby achieving another huge upgrade in speed. We, and many other beetles, stay warm by shivering, so we can move fast without depending on direct sunshine. Thus, I could catch the nonshivering tiger beetles on any cloudy day.

Now, revisiting this childhood place in Germany, I do not at first recognize the forest. The young trees have grown much. But I am surprised at how familiar the road seems. Jogging along, I see at first only a few snapshots of what the forest used to be, but the deeper I go into the forest, the more my mental images merge with the reality of the past. The wood pigeons coo, a jay screams raucously, a raven croaks. The chaffinches and the chiffchaffs sing. After I reach the small sandy road where I ran barefoot to chase tiger beetles and Grützmann's car, I know I'll soon be home.

I barely turn onto the sandy road when I see a large,

blackish carabid beetle running along ahead of me. Strange, because this is one of the nocturnal species that fell into the pits in the ground that Papa used to dig to catch small mammals, whose meat we ate and whose skins Papa sold to museums. I didn't recall ever seeing beetles of this particular species in the daytime, and to see one now present itself to me after all those years almost seems magical. I pick it up, smell its acrid defense secretions, put it back down to let it run on, and then continue to jog on myself.

I'm anxious to round the next curve, past the slight rise where the bees and wasps used to tunnel in the sand and where I once found a nest containing two plump young wood pigeons in the pines. These squab tasted even better than the fried chicken of my imagination. Here is the place where the small gurgling brook paved with black stones passes under the road, where I'd caught red-specked brown trout on their way up to spawn.

Is this really *the* brook? Where is the tiny footpath along mossy banks that led into the forest to the cabin? I suddenly recognize the alder tree where I had found a lichen-camouflaged, baglike nest of a tiny long-tailed tit. I stop. The brook and the places where Marianne found a dead elk and I a wild boar are coming into focus. I've found it! I see the still-faint impression of a footpath into the woods! The hut would be just around the corner, up a slight rise through some beech and pine trees.

When I saw the footpath, I came to a sudden halt. The past flashed over me like a hot breath from nowhere, and it wilted me on the spot. I stumbled, buckled over, and sobbed uncontrollably. I could not stop shuddering for a long time. Maybe I saw a stranger—of the past—on that path, and that stranger was me. But it could have been anyone. And with that second realization I saw kids everywhere, whose fates are determined so profoundly by seemingly insignificant trifles.

When we left the Hahnheide and came to America, in the early spring of 1951, it seemed to me we might as well have been departing on a one-way rocket trip to the moon rather than on a steamboat across the Atlantic. The possibility of ever coming back never occurred to me at that young age. We were alive and had lived from one day to the next. Here we'd found serenity and beauty, after escaping violence and the specter of war. We'd bypassed the mainstream, and I'd got a head start in life because I had learned basics—life cycles of moths, the needs and manners of a baby crow, and the joy of running after tiger beetles through warm sand on bare, tough-soled feet.

FIVE

High School Cross-Country

Be a good animal, true to your animal instincts.
—D. H. LAWRENCE

My mother, barely five feet tall and weighing about a hundred pounds, and my father, also not a natural-born lumberjack, worked together as a team with a crosscut saw that first winter in the Maine woods. The snow that winter (and since) was deeper than anything we'd ever seen. When a tree fell, the trunk got buried and had to be shoveled out by hand, to be sawn into four-foot sections, and then to be dragged to the road with Susie, our neighbor's draft horse. Mamusha and Papa were later even less adept at helping to make kite sticks inside the dark and dusty confines of a tiny woodshop in the town of Wilton. It was not long before Marianne and I ended up in a boarding school for homeless kids while our parents were off for six years collecting museum specimens. They went first to Mexico and then for years were in Angola, Africa.

At Good Will, as it was called then, there were many trails through a forest wilderness of three thousand acres. All the boys worked in the house, the barns, the fields, and the woods. During those six years there, I went from being a "houseboy," washing dishes and floors, to becoming cook, to shoveling manure and helping with the twice-daily milking, to ultimately the top position—runner of the mail route.

In our play at Good Will, we often lived in our own little world of Indians and frontiersmen. The woods were our proving ground for survival skills. Some of us built crude cabins, waged mock wars, and occasionally even killed and ate a porcupine or snowshoe hare. I eventually cleared about a half-mile stretch on one of our forest trails that I ran all by myself, feeling the freedom of wind on my face as the sole but ample reward, visualizing myself as a loincloth-clad Iroquois brave who would be strong and free.

By venturing deep into the woods, we could explore unknown territory at the edge of our tribal world. We armed ourselves with spears made from maple saplings and practiced our throwing skills in remote clearings. Soon after the ice went out of the Kennebec River and the spring sun shone warmly on the yellow matted grass, several of our tight little gang lay naked on the ground in our remote clearing, trying to acquire the proper complexion. In our boyish ways we thrilled with excitement at the thought of fighting other, rival gangs and to hunt and bring back food to our forest campsites.

Philip and Freddie, my two buddies, were at first skeptical about eating common pigeons, which populated the barns with the cattle, sheep, horses, pigs, and chickens. But to me they seemed prime food, like the wood pigeons from the Hahnheide pines. However, eating out is also an adventure where atmosphere matters. For our eating adventures we helped ourselves to one of the school's rowboats that were anchored in a quiet cove of the Kennebec River near the gardens. Then, like the cannibals in the story of Robinson Crusoe, we struck out with our by-then-dead captives, rowing to sandbanks on the far shore, over to an area known as the Pines near a large level clearing of short grass. I loved those sandbanks along the river, because bank swallows had tunneled into them to build their feather-lined nests of grass and laid their pearly white eggs. A pair of belted kingfishers dug their own much larger burrow here as well. The kingfishers' travels into and out of their tunnel left two parallel grooves in the sand as they waddled in and out on their short, stubby legs, which are otherwise useful only for perching on a snag, from which they dive bill-first after minnows.

The grass of this pine-and-birch-enclosed clearing by the river was prairielike, and about a foot or two down from the top of the sod was—still is—a ragged black line in the fine yellow sand. It extended almost the whole length of the hundred-yard-long bank. The charcoal from ancient Indian campfires! At the time that this charcoal was deposited, and in the eons after that, the river must have swollen to much greater height than now, in order to deposit those layers of silt. Maine was partially covered by glaciers, and this was later tundra, populated by caribou and ptarmigan.

I envisioned the Indian encampments along the river of perhaps ten thousand years ago, when the glaciers retreated, and I searched for clues of what they might have been like. There were bits of bone among the charcoal, and once I even found a small stone ax of greenish, smooth-ground stone—a tomahawk (technically a "celt")—among charcoal and three fire-reddened stones that had protruded from the sand. The small, otherwise unremarkable celt tapers smoothly from a 1.6-inch thickness to a 2.6-inch broad cutting face. The ax head made an impression on me mainly because the finely ground cutting edge had at least ten chips apparently knocked out of it at about the same time; there was no wear on any of the chipped edges. Someone had, apparently, deliberately and repeatedly hammered this valuable object against a rock before tossing it into a fireplace. Had two parties of competing hunters met here on the riverbank and, through a symbolic act following a feast of caribou, "buried the hatchet" by destroying it? If so, the reminder of their act had lain through the millennia, to show that our penchant for warfare is as strong as our yearning for peace.

The river flowed smoothly and swiftly along the bank. After we pulled our rowboat onto the sand at the bottom, we secured it onto a snag with a towline. Then we scrambled up onto the grass. Our feeling of remoteness, timelessness, and independence was complete, and we soon drew together around our own campfire among fieldstones. While roasting our plucked squab to a rich brown on fresh maple sticks, we undoubtedly complained about our strictly confining existence on the other side of the river. Tasting freedom was the best part of the meal.

In spring and early summer we'd all spend unbearable hours on our hands and knees, up and down endless assigned rows of vegetables, pulling out the blanket of weeds threatening to choke out everything we planted. But the gardens yielded much more than food. Over the years, the school gardens along the river had produced flint knives and arrowheads. Many of them were then in glass cases on the top floor of the huge brick school museum. The Bates Museum, as it is called, was once a great showcase, but it was then in disrepair and not open to the public as it is now again. It was then inhabited by colonies of little brown bats that squeaked in the walls and left a musty smell. It was a house of treasures, and I was one of the few persons who had reason to enter the museum—through a broken window in the basement. Down there in the dark, not far from the stuffed polar bear, caribou, and bobcat, was a jumble of ancient carriages and farm implements. This was the safest place for me to hide the pupae I'd raised from the tomato hornworms (i.e., hawk or sphinx moth larvae) I'd found while weeding the garden. I overwintered many pupae here, not only those of these large hawk moths.

In late summer, the river in between the school gardens and our safe and lonely camping place on the sandbanks was at times almost solidly covered with pulp logs floating from the north woods down to the paper mills below. Each had a paint mark on the end, to signify specific company ownership. Yet in the coves that were cordoned off with long boom logs strung across the entrances there grew pickerelweed with bright blue flowers. Bass and pickerel lurked under shadowing lily pads, to lunge at the metal "goldfish" and bright red and white "daredevils" that we flung out with fish poles and reeled back, hopeful for a strike. Black ducks nested in the weeds on the wooded banks, and little green herons built ragged stick nests in the dense willow tangles.

Martin Stream, which flows into the river and marked the boundary between the boys' and the girls' farms, was cool and shaded with hemlocks and always free of logs. It had many great swimming holes, but our official swimming hole, the one where we went when supervised, was about a half mile upstream by a large hemlock tree that leaned far out over the water from a high bank. A long rope dangling from it propelled a continuous progression of us naked boys far out onto the stream. We swimmers were always naked, per regulations, because a boy had drowned once after his swimming trunks got entangled on a submerged snag. Nakedness was one of the few rules we willingly obeyed.

A well-worn trail along the banks of Martin Stream led far upstream, past the hermit's cabin and up to the trout fishing holes. I ran there often on Sundays, hurrying to be there as soon as possible right after church to practice my swimming strokes. I found bees streaming in and out of a hemlock tree along this path, and with Mr. Graft, a teacher who also liked bees, we cut down the hemlock and hived the bees. This adventure nourished much longer than the honey. One other time, while scanning for bees in the trees, I spied a tiny owl, no larger than a coffee mug. The yellow eyes of the saw-whet owl looked at me in surprise, as mine looked back in wonder. I needed this creature. I craved it. So I picked clay from the stream bank and with my slingshot, still as in the Hahnheide my most prized possession, I hit and stunned the owl with a clay pellet. It revived soon after I had it in my hand. I could not get enough of the little owl and maneuvered it into a cage in a hiding place up in a spruce tree in the woods. I felt the bird would be at home there and I could see it frequently. But after a few days my memory of it was secure, and I let it go to live free in the woods.

I loved the old sugar maple trees in the woods just behind Guilford Cottage, along Uncle Ed's Road. In May, the pointed leaves of the brown-mottled trout lilies poked through the layers of matted moist maple leaves on the ground, to be followed by bright yellow nodding flowers. Blue and yellow violets, purple and white trilliums, and Dutchmen's-breeches bloomed here as well. One day, while practicing hand-over-hand rope climbing on one of these trees, I heard a faint, dull hammering sound above the usual shrill chatter of the overbird and the lisping of the just-returned yellow-rumped warblers. I tracked it to a nearby maple, and then noticed little chips of rotted wood on the ground. Looking up, I saw a red-breasted nuthatch fly out of a tiny round hole in a dead tree limb about fifty feet up. The nuthatch flew to another branch, shook its head, and released a billful of wood chips, then returned into its hole and resumed the hammering. The nest inside was built of fine strips of cedar and birch bark, and the four fresh eggs were evenly marked with brownish lavender spots. I hadn't yet encountered the mind-set of "Don't swat it, don't catch it, don't eat it, don't touch it" that tends to make nature into a frozen museum exhibit. Nature becomes intensely exciting and real through active participation, beyond just looking at it.

Finding birds' nests in the spring became my burning passion, and through it I observed birds closely, to learn and record intricacies of their habits and habitat requirements, and maybe to "love" them. To be safe, it was best not to reveal that love. As much as I possibly could, I tried to hide my interests, although some of the kids may have suspected them when in our evening study hour and Bible reading I often sketched birds instead.

I'd eventually run away from the school with Philip and Freddie. We walked fifty miles one day and one night before we got hungry and tired enough to be willingly caught. I was much on punishment duty as the housemother made me stay in, first to wash every wall and ceiling in the house, then paint many of them. I had almost no free time, not even to earn money on Saturday afternoon so that I could buy such necessities as clothing and toothpaste. Near that time we took to sneaking out at night. We'd slip out from the otherwise watchful eyes and ears of Mrs. Lizotte, our ultra-strict housemother. We'd leave rumpled clothing as dummies in our beds, then tiptoe down the stairs from our bedroom, but only after we'd hear her impressive snore.

We went in the moonlight to the barns, where Philip worked with the horses in the daytime. He knew what to do. We saddled up, and galloped at top speed across the fields, once even venturing all the way across the Martin Stream bridge to the girls' farm, probably hoping to meet up with other night wanderers.

In the winter we went night-skiing instead, often to the old golf course, then overgrown with brush, to practice our downhill or jumping technique on a snow jump we'd built previously. We also made nocturnal excursions on skis or snowshoes into the woods. It was strictly forbidden and therefore an exciting adventure.

Winter was, next to spring, my favorite time of year. There was then no tedious and endless crawling in the dirt in the hot sun for weeding. Instead, we worked in the woods chopping trees. There were crews of us working together. We built bonfires of brush, and the girls sometimes baked cookies and doughnuts and brought hot chocolate.

At night, as I lay awake on my bed, I could sometimes hear the ice crack down on the river. It was a booming sound, like a combination of thunder and rifle shots, and on cold clear nights that sound carried for miles. In contrast, the silence of gently falling snow was peacefully mesmerizing. Once, in a snowstorm, I came close to a flock of white-winged crossbills. The purple males on the snow-laden fir branches contrasted beautifully with the greenish yellow females.

When Philip, Freddie, and I went on excursion, we wandered mostly to our camp deep in the woods, toward an area called Kendall Annex, an abandoned farm donated to the school. (It would later be sold off as the site for the construction of a giant paper mill with metal stacks that you now see for miles.) It was then all woods and overgrowing fields, and it was the edge of the world as we knew it. We'd eventually push the boundaries when we struck out to try our luck on the aforementioned runaway venture. We didn't even leave the school premises the first two days. It started to rain in torrents, and the crust on the deep snow that April melted, making walking in the woods almost impossible. Martin Stream was quickly flooded and we couldn't cross to Kendall Annex. So we returned along the stream bank and spent the two days under an overturned rowboat in back of President Garrison's house. Meanwhile the state police searched fruitlessly far afield.

We walked across the bridge after dark when the rain stopped, and then we went on into the night to start our long walk. I was cheered somewhat by hearing dogs barking on isolated farms we passed, and by barred owls hooting in the swamps. I had once found a barred owl's nest in the dead top of a large basswood tree, and I'd come back there several evenings to watch and listen to the bird's amazing caterwauling. In college, I wrote one of my many required English papers on that experience. We normally wrote themes about poems, like the one about how "lovely" a tree is because "it has a bird's nest in its hair." I'd never seen, heard, or felt anything like what I'd read in any of those poems, and so I generally got C's and felt thankful. I'd resisted the temptation to write with a pen rather than in pencil because I didn't want to be putting on airs. For once the theme I'd written, on the owl's nest, felt right, but this time the instructor said the writing was out of my style, and "obviously" too good—hence, as far as he was concerned, it was proof that I'd cheated.

Running would be different. What anybody might think does not matter. Credit would go where credit is due.

We had few formal athletics, but Mr. Moody, our eighth-grade teacher, made a long-jump pit next to our big brick grammar school and encouraged practice and competition. I loved the fast sprint, the takeoff from the edge of the pit. Virtually flying through the air, I would land in the soft sand, mark the spot, and measure the distance.

Running speed was basic to gaining momentum and making distance. Running, Mr. Moody told us, was of more significance than just being able to long jump and make a mark on the sand. The Iroquois Confederacy of six tribes that dominated upper New York State had been potent as a result of the speed of its runners, who could quickly carry messages along the 240-mile Iroquois Trail through the wilderness. Relays of runners could cover the length of the trail in three days, and that speed of communication made the confederacy possible. The Iroquoian tribes held running contests, as did the Mandans of North Dakota and numerous other cultures, from the Incas to the Greeks. The Mandans cleared about a 3-mile track in a giant horseshoe, and winners received a red-painted feather as a victory token that could be exchanged for goods.

Mr. Moody redirected our sometimes destructive urges by encouraging us to aspire to athletic skills. Miss Dunham, another grammar school teacher, gave us further inspiration and direction. She told us about Indians who could catch deer by pursuing them relentlessly and wearing them down. The deer I'd seen ran swiftly through the forest in giant leaps and bounds. They represented the ultimate in running performance, and I could not envision them ever tiring. How could a human possibly catch one? She didn't say, but she did tell us about Roger Bannister, who had just three or

Deer

four years earlier run a mile in under four minutes, which had until then been considered physiologically impossible.

We didn't satisfy our urges in real battles or actual hunts, but eventually we found an outlet for all the running we needed in cross-country competition. Bob Colby, our high school English teacher and cross-country coach, had announced at the morning convocation in early fall of my junior year that all boys who were interested in running cross-country should meet him in his homeroom after school. Cross-country is a fall sport, and at Good Will, with fewer than a hundred kids in the whole high school, it was, next to baseball, a major sport.

"Cross-country is a team sport," Mr. Colby began after we assembled, "where the team with the lowest score wins. The first man in gets one point, the second gets two points, and the twentieth gets twenty points, and so on. Five men score. A team that gets these five men in first scores one plus two plus three plus four plus five—equals fifteen points. That's a perfect score. Those runners who don't score directly can still help the team win by beating other scorers, so that their numbers are higher. Any questions so far?"

The rules were simple and direct. The better anyone on the team did, the better for all. I liked the idea of individual initiative contributing to our team of Harriers. Here was my chance to belong, but could I make it? As I looked around, I wasn't overly confident. There was Jerry, who was already man enough to shave. There were about twenty other hopefuls. At the age of nineteen years then, I was the smallest, most peach-faced, and underdeveloped of the lot, with only a few pimples showing a first sign of adolescence. Worse, I was called Nature Boy because I liked bugs and birds. There was Jughead, a foot taller than I. He was thin and lean and had his long, dark hair slicked back with greasy Vitalis. "Did you ever notice," he asked Mr. Colby, "that intellectuals aren't developed physically?" I did not by any means claim or want to be an intellectual. Nevertheless, the assumption behind that question—namely, that physical excellence might just possibly exclude thoughtful pursuits, maybe even loving birds and bugs—agitated me. It was too much in line with other forced stereotypes. "No," Mr. Colby said, "I haven't noticed."

Our educational programs incorporate physical training, assuming a mind-body connection. Plato, who participated in the Isthmian Games as a wrestler, as well as Socrates, who was said to keep himself in excellent condition by training in a gymnasium, emphasized the necessity of physical training in a sound education. Some of Plato's dialogues, I later learned, took place literally in a gymnasium. The Indian and the Greek ideals that we have adopted emphasize the development of the whole person.

"We'll train for two weeks, and then we'll have time trials to see who makes the first string—that's the first seven. Those seven who make the team get to go on road trips," Mr. Colby continued.

"Road trips," Mr. Colby went on, "are when we go to race at other schools." Our longest road trip would be to Vinalhaven, an island off the Maine coast. We'd take a ferry ride in addition to a long car ride. I wanted to make the team more than ever. But could I outrun enough of these brutes to do so?

We were issued uniforms and equipment. We each got one pair of white cotton socks, one jockstrap, and a pair of narrow black canvas running shoes with a thin, hard rubber sole. Our running uniforms were black shorts and an orange T-shirt with the Good Will emblem. We met on the lawn outside for our first training session. We met there every day, at 3 P.M. sharp. As we lined up on the lawn in back of Averill High, Coach Colby began our routine by having us hop while swinging our arms and legs to the sides. We got down on our hands and in unison pushed our bodies up and down from the grass.

Among the Penobscot tribe in Maine, each family group used to have some young men who were specifically designated as runners to chase down moose and deer. These "pure men" were chosen because of their fleetness of foot, and to be one of these few was considered an honor. The runners were guarded by some old men to make sure that they did not have sex, that they slept with their legs drawn up, and that they did not chew spruce gum, as those transgressions were thought to impair their breathing as well as make their testicles clack when they ran, warning the deer. Change only a few of the parameters, and we students at Good Will were like those pure men. We also were fed myths that, from a more rational (that is, distant), unbiased perspective, were probably similarly absurd. I never did figure out why we all had to wear jockstraps, for example. But I would never have dreamed of going without one.

Putting on uniforms and doing jumping jacks and push-ups in unison does not make a team. That takes pride, and pride comes from exclusivity. After the warm-up, our real workout began. We were told to run up Green Road, turn left at the top and come down Uncle Ed's Road, run down past the front of the cottages, and then finish up in front of Averill High. That would also be our home course for meets. The whole distance was close to 3 miles, and it was almost all on dirt road. "Now let's line up here . . . Ready, set, go."

We were off in a rush, Jughead and Jerry galloping on ahead. Everybody tried to keep up with everybody else. It seemed forever by the time we finally all made it back to the school building, but I was glad I was not one of the stragglers. Some weeks we probably ran close to fifteen miles. The consistent stragglers soon handed in their uniforms. We were beginning to have a team.

Our first meet during the next year—my senior year—was a home meet. I was "going steady" at the time, which basically meant I had butterflies in my stomach when I happened to meet the object of my affections in the hall. I very much hoped to see her at the finish line. Imagining her there, I counted the days until Friday afternoon, the time of the first big meet.

I remember the tension beforehand, and then the lonely, agonizing battle of about eighteen to twenty eternal minutes as I strained every nerve and muscle and never once looked back. My overpowering wish was that it might end. During it all, I anticipated how good it would feel finally to stop running and to see my girlfriend's face. Coming across the small cement bridge through the pines, I could see the finish line, where several spectators had gathered. I rushed to it and nearly collapsed in ecstasy. I was first. As is usual in situations like that, the pain is soon forgotten. The happiness remains.

My surprising ending seemed like a fluke, but it then happened four more times in a row. After that, I was no longer derisively called Nature Boy. I was instead "an animal," which of course we all are. However, that sounded much better. In fact, it felt great.

Every morning before classes, all of the students and teachers gathered together in the school auditorium. In unison we held our right hands over our hearts and solemnly chanted the Pledge of Allegiance to the flag and then sang a patriotic song about bombs exploding in the air, which made me cringe. After bowing and murmuring for a while, we lifted our heads and listened to announcements. On one particular morning after our fifth cross-country meet, Principal Kelly announced to all the assembled crowd that our cross-country team had won all our meets so far, and that I had won the distinction of being the school's first "ace"—a five-time winner. I did not think of myself as a better runner. I had just tried harder and been more careful about what I ate.

I was never known to be a fussy eater, which was sometimes a topic for jokes. However, being as active an animal as I was, I knew proper nutrition was essential to my running. My gums were bleeding a lot, a sign of scurvy, vitamin C deficiency. That worried me a lot.

In one letter to Mamusha and Papa in Africa, dated July 3, 1958, I wrote: "My coach thinks I could become state champion in the fall. . . . Tomorrow I have to go to the dentist, again. He sais [*sic*] he might have to pull a couple of my back teeth—too bad. I don't think it's because I don't brush my teeth enough because I do quite regularly. It must be the diet here—it isn't very good. Many of the other kids have the same trouble. It's no use crying over spilled teeth."

On February 3, 1959, I wrote: "I have a boil right now—we have had an epidemic going round. It is baffling and kind of scary. Yesterday in church Marianne had to throw up twice and one of the girls just collapsed. . . . Quite a few of the others besides myself have boils, too. This all makes me quite made [*sic*] because I think a lot of this can be avoided if they made an effort to provide a balanced diet. At least I think there would be more resistance."

I did my best to provide myself a balanced diet. Aside from the occasional captured squab from the barns in springtime that we roasted out on the flats across the river, I once stole canned fruit cocktail assigned to our housemother, and when I worked at the barns I often reached into the bins and ate the grain mixture meant for the cows, assuming the diet of the cows to be unrefined and nutritious.

My appetite was increased by my new job as mail boy. The mail boy picked up the morning outgoing mail in a leather mail pouch in the Prescott (administration) building after breakfast, bicycled with it to the Hinckley post office, then brought back the morning incoming mail. After school, he did the same again. I left the bike at Prescott, and I ran.

The one-room post office was the domain of Gordon Gould, a short, battle-scarred tank of an Irishman. He liked to be known as Lefty. And Lefty he was to me, and in my memory will remain so always, even though I never actually saw him swing that left hook he said he was famous for when he was on his way to become the welterweight champion of the world, before he got shot up in the war. "I never got knocked out," he told me, and, "I'd run five miles every day, and I could do two hundred push-ups nothing flat."

Lefty was once a Good Will day student, and I felt comfortable with him because he knew what our life was like, and the outside world as well. He had served in the A Company, 504, in the U.S. Army Eighty-second Airborne Paratrooper Division, in World War II, before he came back home to Hinckley crippled, to become postmaster. He was usually the only person I saw in the one-room post office, every morning before and every afternoon after school. The faster I ran, the longer I could stay and listen to his adventures.

Before my two years as mail boy were up, I'd listened to Lefty for hundreds of hours, standing awestruck in front of his little barred window as he talked about his war experiences in Anzio, Northern Africa, Sicily, Belgium, and Germany. I could almost smell the powder, hear the thunder, and see the tracer bullets as he told of his exploits with his buddies Eddie Adamzic and T. J. McCarthy. Beads of sweat would sometimes form on his broad forehead as his gray-blue eyes looked deep into mine, as he tapped his vivid memories.

"One time when we were firing at the Germans on the hillside opposite us, one machine gunner over there didn't hide his position. He'd keep hoisting up Maggie's drawers (a little white flag), telling us we'd missed. When it came time for us to finally advance, we went around him. Another time a couple of them came over at night. They'd somehow got through our lines and completely surprised us—held us up with machine guns, demanded cigarettes, sat down and smoked, and talked with us, and then went back. One night I was inspecting our machine gun positions—and at one the gun was gone. 'What happened?' I asked the crew. One said:

'Well, one of the guys has been bringing up coffee every night from the rear, and to identify himself he says, "Don't shoot—coffee coming up." This night the same thing happened, only the Krauts saying it, and then they told us, "We just want your guns. You can stay. Explain *that* to your officers."

"Heidelbrink (another of Lefty's comrades) could speak German. He went to school there, and knew their psychology. He'd go behind *their* lines. One time he came back wearing a German major's uniform and he had a whole company of prisoners with him. He'd ordered them all to line up and march. They did. They couldn't disobey an order.

"We sang 'Oh Susannah.' Then one night one of them comes over with his hands up, saying, 'I'm not surrendering—I just *love* to sing, and I want to sing along with you guys.' He did, too, in a deep, baritone voice."

Of course, it usually wasn't like this. The officers on both sides kept them moving around, so that they'd remain enemies.

I didn't hear about all the battles, but he did recount riveting events of the last one.

"I saw the tracer bullets coming, then the Krauts had got me—I saw this thighbone lying next to me. I realized it was mine. I got *mad*—I flung it at them. Then I passed out. They overran us. It was kids, really, that dragged me off afterward."

It had been near the end of the war, and he was put into a Belgian hospital. "The German doctor told me that when the war is over, when your army doctors get to you, they'll tell you your leg will have to be amputated. It's the easiest thing to do. Refuse. Your leg *can* be saved."

And that's how it turned out. "When they shipped me stateside to the veterans' hospital, the first thing they told me was, 'We'll have to take off your leg.' I said, 'No.' They said, 'If we don't, then you'll die.' So I said, 'I'll die, then.' " Lefty didn't die just then, but he could never again run.

To me, Lefty was a great friend and ego booster, and if I ran to please anyone, to make them proud of me at Good Will, it was for him, and for Coach Bob Colby.

Our totem at Good Will was the beaver. "The beaver," we were told, "works when he works, plays when he plays, is strong in individual effort, yet labors for the community good." The beaver cuts trees individually, yet its dams and lodges are built and maintained communally by the whole clan. Efforts from one generation of beavers contribute also to the well-being of future generations. This was not just school propaganda. They are ideals that encapsulate what makes us human. We are also social animals, and that sociality has been handed down to us from our apelike ancestors of millions of years ago, as it has in beavers, ants, chimps, and bees. Like other animals, we play at those things that are important to our survival, and social play promotes social cohesion. Our school sports teams gave us a feeling of belonging, an identity as a group. Gangs that fight one another with knives and guns; like those some of the boys I knew had come from, do the same thing, but at a high cost. If we can't find allies in one context, we will in another. But there is a prerequisite: in order to forge alliances, we first need worthy adversaries. Without adversaries, no alliances are necessary.

One morning at the post office, Lefty pointed one out to me. With his stubby fingers he poked into our newspaper, the *Waterville Sentinel,* at a headline about Bert Hawkins, an undefeated cross-country runner from Waterville High who set a course record every time he ran. Hawkins immediately loomed, almost menacingly, larger than life.

There is nothing that can make one feel smaller than seeing someone big, which is why many try to talk down those who are more capable than they are. In running, you can't deceive yourself or anyone else. You have to confront facts; I knew that Hawkins could outrun God.

A meeting with Hawkins was inevitable, since Waterville was just a few miles down the Kennebec River. Waterville High was a class L (large) school, whereas we were S (small). Nevertheless, Coach Colby invited them down to run against us, and they came. I did not see the Waterville Warriors until they came out of the locker room for the showdown in front of Averill High. We were not favored. I knew, too, that in a very few minutes I'd be exposed: I was not a great runner. I just tried harder.

As with all insecure kids, a large part of my existence teetered precariously on a thin, fine edge, with independence on one side and pleasing my all-powerful parents and parent figures on the other. The scales were uneven; my housemother saw me as fundamentally flawed from the start. She called me a little Hun because I had the wrong accent and spoke poor English. Her logic then filled in, supplying an ironic twist to every innocent act of fun, curiosity, and survival, to color them into evil and grotesque crimes. After a few years I felt robbed of precisely those qualities I valued and aspired to. With nothing more left to lose, all I had left to develop for redeeming my pride were dare-devil acts and physical prowess, and I tried a little of both. The first induced me to commit outrageous acts that got me kicked out of school just a week before I would have received my high school diploma, while the second helped me get an education. This race against Hawkins indirectly contributed to the latter.

Someone pointed him out to me. He was the skinny kid

with the black crew cut—the one who gave me just the barest hint of a smile (or was it a sneer?) as we lined up for the start.

As was his well-known custom, Hawkins took off fast and established a big lead. We approached the mile-long gentle uphill grade of the Green Road, where I'd once tried to outrun Principal Kelly in his paneled station wagon after he had caught sight of me during lunch break, just as I was lighting a firecracker that I'd made in chemistry class. The firecracker had fizzled, but it ignited the rumor that "the little German kid was trying to blow up the bridge." But Lefty only laughed at this nonsense and talked to me as always.

This was home turf. I knew that if I could just hang on to the top of the hill, Hawkins might feel as if he were having a bad day. It was my only chance. Gradually I crept up on him—he must have begun to hear footsteps, because he turned around and looked, and before running another hundred yards, to my utter amazement he stopped to calmly take a leak beside the road.

Once you turn on the spigot you can't very well immediately turn it off again; you've got to let it run awhile. I took the advantage and went past to grab the lead. After reaching the top of the hill, it was all downhill, and I pressed my advantage all the way home. By the time I was coming over the little cement bridge where I'd set off my little sizzler, he still hadn't caught up. I could hear the girls cheer and the coach holler, "Atta go, Ben!" (Ben was my nickname throughout high school and college in Maine.) I gathered my last bit of strength and managed to finish just a few strides ahead. Running prowess may seem unimportant to an antelope until that rare moment in its life when a lion gives chase. To me, that last moment had been important. It had been mind over matter.

Modern biology has now proven the physical reality of the mind-body connection and uncovered some of the mechanisms that back then would have seemed like science fiction. The mind serves as the mediator between sensory input and physiological output. Who could have imagined that on a specific schedule of darkness and light, or even at a mere flash of light at precisely the right time, a moth pupa would "decide" whether to remain in torpor for months or to metamorphose into a flying adult? Who could have imagined that a male sparrow, experiencing a longer day length, would get surges of testosterone into his bloodstream that initiate a cascade of physiological changes that then alter his behavior and also cause him to molt his drab feathers into brilliant ones? Who could have imagined that a dove would go through all of the profound physiological changes in enlarging her ovaries and developing and laying eggs, just by seeing some twigs and a courting male? In all three cases, the sensory stimuli excite, or activate, the brain, and the brain induces a cascade of hormones that then affect the body. We humans have the same mind-hormone axis, and we are additionally blessed with consciousness, which can excite our brains sometimes with very little provocation from sensory input; we can intensify any input with the lens of our minds. But there are limits. We can't cure a cancer with good thoughts, but good thoughts can make us feel better and allow us to function more effectively. They can also help us accomplish what seems otherwise not possible.

High school cross-country had provided the transition from running to racing. I had tasted the lure of the chase, and I was changed. As cross-country runners, we learned to focus our energies, to shackle every atom of our energies for one clean task, if only briefly. Yet we worked hard for a specific and well-defined goal not only during the chase but also in long preparation for it. It turned out that for me there was another bonus.

The possibility of going to college had at first been too remote for me even to consider. We had no classes in any branch of biology, and I had done poorly in Latin, which Papa said is the language of biology. How could there be hope for me to do anything in a subject where I did not even speak the language? Chemistry? Instead of doing experiments, we sometimes made firecrackers, but unsupervised. In physics we only read aloud from the book. I retained little from the textbooks, learning instead from what I lived and the things I touched that held emotional content for me. My head was filled with what was (and maybe still is) considered in many circles as esoteric and probably trivial knowledge about almost everything that flew or crawled or swam. I was also influenced by equally nonacademic adventure stories of explorers in Africa. I had absolutely no means of financial support. The few Saturday afternoons I'd had off from work I hired out at the barns for a dollar the afternoon. It was enough to buy my secondhand clothes and other necessities. What more could I want? Nevertheless, consider going to college I finally did, in the late fall of my senior year, when Mr. Kelly, our principal, said to me, "Ben, they have a great cross-country team at the University of Maine." I knew right then that I needed a college education.

Six

College Cauldron

On what wings dare he aspire? What the hand dare seize the fire?
—William Blake

"I want to go out for cross-country," I told Edmund Styrna when I walked into his office in the Athletics Building at the University of Maine at Orono during freshman orientation week. Coach—as I would call him for years after—is a tall, crew-cut man with bushy eyebrows. He beamed broadly, and soon took me to the stockroom, where I was issued running duds. Then he took me to the locker room, where all athletes had their own personally assigned lockers. Such attention from the track and cross-country coach of the greatest running team in the state, at the greatest awesomely big university I could imagine, made me want to run my heart out for him.

All the new, strange sights and smells, the excitement

and anticipation, were intoxicating. On that very day, the first one of my college career, I changed into the new clean clothes, ran the cross-country course in my boots, and followed up with a workout in the weight room.

I didn't have a high school diploma, but that didn't matter. I had had high school and was admitted before graduation ceremonies in June. During the preceding summer, I had been lucky to get a job working in northern Maine for the USDA surveying for gypsy moths, a notorious forest pest. I'd rented a small room just beyond the "one-hundred-mile woods" in the town of Houlton. Having just barely learned to drive, I headed out every morning in my new government-issued truck, on some isolated road (they were all isolated) to set out or to check my traps, baited with the scent (pheromone) of female moths to trap males of the species. I was alone all day, every day, for two months, except for occasional weekends when I hitchhiked the 180 miles back home to my parents' farm. I caught not a single gypsy moth all summer, basically proving that the moths had not invaded northern Maine. There would therefore be no spraying of biocides.

There was no time left to run after work to prepare for college, so I did it while on the job. After a few minutes of driving to another of my approximately five hundred pheromone traps, I'd stop some 50 to 100 yards from the trap in order to get a short run in. After I ran back to the truck, I'd drive off. Usually, after catching my breath, I'd start singing at the top of my lungs till I came to the next stop. By fall, when I came to the university, I still couldn't carry a tune, but I could run a little faster and farther. More than anything else in the world, I wanted to make the team.

When I went to the weight room on my first day at UMO, I had seen a brawny specimen of a guy lifting barbells. I'd lifted grain bags at the Good Will School barns, where I barely managed to lift a hundred-pound bag over my head, but only by keeping my back straight. The brute I was watching bent over in a jackknife position, and then he lifted while his back was horizontal to the floor. I'd never watched real weight lifting before, and I didn't know how it was done. Like that? Well, I'd do no less. I grabbed a set of heavy barbells, bent over, and started pumping. It wasn't long before I felt a pain in the small of my back. I wasn't going to let pain stop me, so I kept pumping.

In subsequent weeks, I continued to work on the dish crew in one of the cafeteria kitchens because I had no scholarship money and didn't get a penny from my parents—they had less to spare than I did. My back pain persisted, but I still managed to run in slow agony, sometimes with shooting pains going down my leg. The doctor in the infirmary had told me that it was just a muscular strain and would go away. Weeks went by, and it was still there, and then it got worse. So Coach told me to go back to see the head physician. I was supposed to tell him who'd sent me and that I was on the cross-country team.

This time I got checked out more carefully. So carefully, in fact, that I got worried. Finally the doctor told me that I was done running for good. No more work lifting trays of dishes, either. He said something about a ruptured lumbar disc—extrusion on the sciatic nerve. "It can be operated on, but it doesn't really heal up." I was referred to a neurologist in Bangor, who was no less discouraging. He told me that running was out of the question and I should reconsider my intended career in forestry.

I could not run. I was out of a job and therefore maybe out of college and a career. The trail had gone cold, and I'd have to find another. In a situation like that, one can rely on chance and take advantage of immediate opportunities, or one can plan a strategy. Undoubtedly, given life's vagaries and complexities, the latter can be as unpredictable as the former. I had no plan. Instead, I studied as never before and eased the pain somewhat by putting a stiff board under my mattress. My emphasis on studies was a good thing, too, because like my high school classmates I was unprepared for the intellectual rigors of college chemistry, calculus, and physics. As far as I know, none of them made it beyond freshman year. The professors did their best to weed us out, to maintain high academic standards. As it was, my back injury and subsequently also a knee injury kept me bound to the books and thwarted their efforts. These injuries had another consequence for me as well. It had to do with the military.

After I got my running duds, the next thing I had been issued at the university was a U.S. Army uniform. All the men were required to take the army's Reserve Officers' Training Corps (ROTC) course for two years. It was just part of the deal if you happened to be a male at a state university. All of us freshmen and sophomores shined our black shoes and the visor in our flat-topped olive hat to mirror sheen once a week. We all showed up for weekly drill in full dress uniform at the field house, to assemble and march in squads, platoons, and companies as army officers hovered all around with watchful eyes. Upperclassmen who'd been promoted yelled commands as we plebes paraded around, practicing staying in step and in formation, carrying and presenting our bulky M-1 rifles. These sessions were agony on my back because I could not bend to loosen the strain on my ruptured lumbar disc. I did not want to have to leave the university, so I did not complain but endured my two years of obligatory ROTC service. We all owed Uncle Sam real military service afterward. The only question was precisely when and where. I didn't want military service hanging over my head after college for too long; I wanted to get it out of the way as soon as possible. So, just before finishing my fourth year of undergraduate studies, I went to see the army recruiter in Bangor.

I passed all the tests. I was well qualified because, as I'll explain later, I had in the meantime recovered and also had a whole year of jungle immersion in East Africa. The recruiter thought I'd be an ideal candidate for some jungle country in Asia, where they'd need me for something other than shooting birds. I was a good rifle shot. When I couldn't run I often went to the ROTC rifle range for recreation because Sergeant Bell, who was in charge, let me shoot all the ammunition I wanted. He said I was a "natural," and he wanted me for the rifle team. The .22 rifle that I'd earned by doing chores with Phil Potter back home had been my most prized possession, and Phil and I had done a lot of target practice. Also my pulse at rest was slow enough so that I had time to squeeze off rounds in between heartbeats to reduce the scatter of shots about the bull's-eye.

All was going well. The recruiter smiled. He said I had all the credentials for my first choice, the paratroopers. All I needed was a letter from my doctor since, as requested, I had admitted to my former injuries. No problem. I went to see Dr. Graves, our campus physician, who'd seen me a lot already. "Doc," I said, "I need a letter for the army." "OK," he said, "you can pick it up tomorrow." The next day he had written a letter and sealed it. When I went to pick it up, he handed it to me and said, "This ought to fix you up, Ben."

It seemed like an odd comment. But I brought the letter straightaway back to the recruiter, who took it from my hand and opened the envelope. Then he turned his back to me, and as he was walking to see someone in the other room

I heard him swearing. Puzzled, I left. Soon after that I received a new draft card in the mail. Instead of the 1A classification I had since I was eighteen years old—when I registered for the draft after I'd been sworn in to become a U.S. citizen and was duty-bound to defend the country—I was now 4F: I was now deemed physically unfit for military service. That's why I ended up in Dr. Dick Cook's lab, washing glassware and doing other routine chores before I finally worked on the mechanisms of cell respiratory physiology. That's why I didn't jump out of airplanes at the parachute training center at Fort Benning, Georgia, or elsewhere, as Lefty had done.

Although I had sustained a major back injury during my first week in Orono, I did eventually become a member of track and field and cross-country teams. I was not big like many of the field men, nor did I sprint like a gazelle. For too long I had compared myself to others in terms of size and strength and speed, and invariably found myself not up to par. But here, on the track and field teams, a diversity of skills could be found: hurdlers, sprinters, hammer throwers, javelin throwers, shot-putters, middle-distance runners, high jumpers, long jumpers, and pole vaulters. I became a distance runner. I did not have to be like anyone else, an important lesson for life.

Distance runners have one common trait—the good ones are skinny. Weight specialists such as shot-putters or hammer throwers are a completely different sort of animal from distance runners. The two represent extremes of body build, coordination, speed, and endurance. Numerous and diverse aspects of physiology underlie those differences. The weight specialists, in order to be at the far extreme of what is humanly possible, must have a massive body of bulging muscles and thick, strong bones to support them. They require a high percentage of fast-twitch muscle fibers that anaerobically (without oxygen) burn carbohydrate for explosive release of energy. Their competitive event generally takes a second or two; the preparation, years.

The distance runner must fairly float along the ground, sometimes for hours on end. Ideally, he has light, thin bones and long, thinly muscled limbs, like a bird. The key to the distance runner's performance is to supply his fat-burning muscles with a sustained supply of oxygen. That capacity involves a large support system that includes a large heart capable of a large stroke volume per beat, rapid beating if need be, and slow if not. He needs large arteries, extensive development of the capillaries, large lung capacity, and large fuel depots in the muscles, the liver, and other areas of the body. His cells must be packed with mitochondria, the microscopic power units that, with their batteries of enzymes, convert the fuel and oxygen to energy, which is then harnessed for muscle contraction. The quick power of the sprinter or thrower requires no mitochondria and hence no oxygen, or the support systems for oxygen delivery.

The body's ability to deliver a continuous supply of oxygen to the muscles (as well as to the brain and to all other organs) is put to the supreme test in distance running. The lung-heart, or bellows-pump, mechanism is extremely important in this task, but the blood is even more so. Our blood is highly specialized for ferrying oxygen molecules from the lungs to the mitochondria, acting in concert with short-range-transport mechanisms across membranes at the muscle cells themselves.

The oxygen-ferrying capacity of blood pumped by the heart is boosted nearly a hundredfold above that of the plasma by containing oxygen-carrying vehicles, the red blood cells. Each red cell of our 25 trillion blood cells is packed with millions of iron-containing protein molecules, called hemoglobin, and each hemoglobin molecule can load up with four oxygen molecules in the lungs and then upload them to the capillaries, such as those in the muscles. Hemoglobin is called respiratory pigment because it has a bright red color when it is loaded up with oxygen, though it turns blue when the oxygen is unloaded, as when the blood in the veins makes its return trip back to the heart and lungs.

As oxygen continues to be transferred to and dumped off in the capillaries, it builds up large concentrations that prevent further hemoglobin from unloading—and thus make the heart's work of pumping blood a wasted effort—were it not for a second pigment that is very similar to hemoglobin. This protein, called *myo*globin ("myo" refers to muscle), is what makes meat red. Myoglobin is within the muscle fibers (cells), and by binding oxygen there even more easily than hemoglobin binds oxygen in the blood, it removes oxygen from the blood and makes it available to the cells' metabolic pathways. Oxygen then follows its gradient of concentration, from high in the blood to low in the cells, where it is being used up.

Not all meat has myoglobin. As we all know, chicken has white meat as well as dark meat, which ensures that at every picnic there is debate as to who gets the white (breast) and who gets the dark (leg) meat. (I generally choose the dark meat because of the iron in the myoglobin, which all runners need.) White meat is composed primarily of the fast-twitch, anaerobic muscle fibers, capable of explosive, or sprint, power; and red meat has a preponderance of slow-twitch, oxygen-requiring fibers, of less powerful contractions but great endurance. A grouse has white breast muscle like a chicken, and it seems to virtually explode in a burst of power when it shoots up in flight, like a firecracker going off. It can't keep it up, though. After a few such flight explosions in succession, a grouse is rendered flightless. On the other hand, it can run forever using its dark leg muscles. Long-distance fliers like long-distance runners require dark meat, and migrants such as warblers, sandpipers, and geese all have very dark breast meat (that is, wing) muscles.

We have both fast- and slow-twitch muscle fibers in the *same* muscles in our legs, and the mix makes our muscles appear neither white nor dark red, but presumably an intermediate pink. The distance runner's leg muscles contain predominantly—79 to 95 percent in elite distance runners, versus 50 percent in the average individual and about 25 percent for elite sprinters—the slow-twitch variety, which burns fat and requires a continuous supply of oxygen to operate, so that it doesn't leave lactic acid, which quickly causes fatigue.

Different individuals have different percentages of fast-twitch as opposed to slow-twitch fibers, which predisposes them to either sprint or endurance performance. It is thought that we are born with our specific muscle-fiber-type ratios. However, no studies are available that have followed the muscle fiber type of an individual toddler to an adult sprinter or a long-distance runner; we don't know whether muscle fiber type is predetermined at birth or becomes fixed at some early age due to one's lifestyle.

Researchers determine fiber type percentages (relatively painlessly, I've been told) by snipping biopsy samples directly out of our muscles and then staining and microscopically counting the different kinds of fibers, to determine one's potential for sprint versus endurance events. To some extent, fiber type changes with training. More recently it has been recognized that there are two types of fast-twitch fibers, termed a and b. The FTa fibers are a bit more aerobic than the b, and they become so with training. The average person's 50 percent of fast-twitch fibers are equally allocated between a and b, but elite marathoners end up having almost no FTb fibers. Fiber type is thought to be determined by the nerves innervating them. One neuron activates numerous fibers all at once in what is called a motor unit. Fast-twitch motor units typically involve one neuron innervating three hundred to eight hundred fiber cells, whereas slow-twitch motor units consist of ten to one hundred fiber cells. Training involves not only the biochemical adaptation of the fibers, but also neural coordination in recruiting them for their work.

By having both types of muscle fibers in one muscle we gain flexibility. We get both power and endurance, but such flexibility is a compromise. An elite sprinter necessarily loses endurance, and an elite distance runner loses explosive power. Why, then, must a muscle that is built for endurance sacrifice power? The answer is probably related to the fact that for many repetitive contractions to be possible, valuable space otherwise available for muscle fibers must be sacrificed for mitochondria, for an extensive blood capillary net, for membranes, and for myoglobin, and of course for the pump-bellows support system. An anaerobic fast-twitch muscle needs no provisions for an immediate supply of oxygen or for fuel, waste disposal, and temperature regulation. It's like a race car that's designed to go very fast once around the track, as opposed to a camper that must contain all the provisions for crossing a desert.

Many of the factors that combine to produce sustained power in the middle-to-long-distance runner are proximally measured by the capacity to process large volumes of oxygen for making the aerobic metabolism at the level of the cells possible. The *maximum* rate at which we can process oxygen depends on all of the many variables we've just considered, and many more besides. That maximum volume of oxygen that we can process on a sustained and steady basis over a matter of minutes is abbreviated $\dot{V}_{O2}$ max. And $\dot{V}_{O2}$ max must obviously be quite high in any distance runner who wants to have a chance at getting into the big leagues.

But $\dot{V}_{O2}$ max is not the limiting factor of power for sustained durations. When the fuel runs out, even a Ferrari is forced to stop. Unlike a Ferrari, we're warned long in advance of impending fuel depletion—we start to feel weak and are forced to slow down long before we're empty. If his specialty is to run very long distances, then the runner will also be tested for his ability simultaneously to mobilize fuels from body depots and to operate his gut to supply fuel on the run.

Indirect physiological steps are critical as well. A distance runner generates prodigious amounts of metabolic heat. He must be able to dissipate that heat by sweating, which requires intricate processes of salt and water balance and decisions of routing blood either to the skin to get rid of heat or to the digestive system or to body fuel depots. His kidneys and liver must continue to operate and process and eliminate metabolic wastes. A weight lifter, jumper, thrower, or short-distance sprinter has more than enough fuel right in his muscle cells, and since he needs no oxygen during his event and has no stresses for heat dissipation or waste elimination, he can essentially put most of the body's systems on hold, and concentrate on explosive power. His task is the quick release of massive amounts of energy from a preponderance of fast-twitch muscles, harnessed by a superb biomechanical capacity and learned coordination in speed and agility.

The runner is also tested for biomechanical efficiency, especially on longer runs, where fuel and energy economy become crucial. All of his motions must be harmonious and seamlessly choreographed in a fine-tuned coordination of hundreds of muscles and thousands of muscle units, for one integrated task, one huge reflex mechanism. Whether or not he consciously learned to do so, his arm swings are precisely integrated with leg swings, or stride. Both are meshed with his breathing and likely with his heartbeats as well. At the most efficient running stride, arms, breaths, and heartbeats are multiples of one another. Those multiples change with pace and effort, but the synchronicity does not. It is as though his legs beat the tune to create the body's rhythm. However, there is an ever-changing rhythm throughout the race as strategy and tactics change in response to current progress and competitors.

In competitions, we test seemingly simple tasks that any fit child can perform, namely running, jumping, and throwing. Their beauty reduces to the ideal of coherence and simplicity; all that is essential is present in precisely the right proportion for the one circumscribed task, and nothing is superfluous. In college field and track, I saw extremes of these specializations. I saw what a lot of us have, but raised to higher levels. I saw the product of dreaming and our indomitable drive for perfection.

The steady improvement in world records of all sporting events during the last hundred years (when records have been kept) may look like biological evolution. We don't know *all* the factors that have been and are involved and to what extent they have had relative effect, but there *is* one thing we can be sure of: it's not due to biological evolution. Evolution might still have played a role shaping us back in the ice ages, when we were fragmented into small, isolated populations and faced death regularly due to athletic deficiencies. No longer. We're now in large, ever-more homogeneous populations where any mutation that could potentially be of value for athletic performance, in the context of a specific suite of characteristics, will be quickly swamped. Nor do we face directional selection in favor of jumping, running, or lifting ability.

It is tempting to extrapolate athletic records to the future and assume they will improve forever. The obvious, trivial truth is, they won't. In just one century, the law of diminishing returns has already set in. Despite vast improvements in nutrition, a quadrupling of world population, a skyrocketing participation (at first involving just a few mad Englishmen) of athletes from almost the whole world and from most of its social classes; despite revolutionizing training techniques based on vast data banks of scientific research; despite full-time athletes, backed by specialty state-of-the-art Olympic training centers; despite vast improvements in equipment (shoes, poles, javelins, synthetic tracks, waveless pools, aerodynamic and hydrodynamic clothing), we are now seeing decades when records are broken by only fractions of seconds. The bar of acceptable international performance has been raised very high. There are still part-time college athletes who dream of Olympic glory, as some of us once did, but there is no longer any chance of being the best out of a pool of 6 billion people by part-time participation. The best today will now no longer be someone who flips hamburgers eight hours a day and trains evenings or weekends.

Jim Thorpe, perhaps one of the best athletes of this century, had his Olympic medals stripped from him because he accepted petty cash in a minor baseball game. Today all Olympic athletes are supported in some way, usually *all* the way. Performance-enhancing drugs to build muscle mass are more than suspected in some sprint and weight events, where bulky muscle mass is an advantage rather than a hindrance. Indeed, almost any freak performance (that is, far above the usual—as is now necessary for almost any entry to world-competitive status) is now suspect almost as a matter of course. In my opinion, that mind-set alone will kill the onward march of record performances, and the college athletics that spawn them, dead in their tracks, even before we exhaust our biological potential. Why? Because runners compete for glory. There is no glory in drugs. If drugs are *automatically* suspected, then few except the professionals will put in the necessary heroic efforts now required; they will not put in all the work if they might, therefore, be accused of cheating!

Some reports claim that withdrawing an athlete's blood and later returning the red blood cells to boost blood volume or the red blood cell count (that is, blood doping) increases aerobic capacity and distance-running performance. Besides the ethical issues raised, the results of such studies are in conflict, and there are risks in performing such a procedure. There are also questionable physiological assumptions being made. It is important to realize that when we lose blood, our bodies respond by replacing that blood. That is, we produce blood *to the extent that the body demands*. Thus, I suspect blood doping may work to temporarily give a boost to an untrained athlete, but it is difficult to imagine that it would not harm a trained athlete in whom all functions are raised to an intricate balance for a specific task. To an elite distance runner, who typically has a higher blood volume but a *lower* red blood cell count than an untrained person (to reduce blood viscosity and reduce strain on the heart), any extra red blood cells can only be a detriment, and artificially adding blood seems a bit like adding a third leg when two are already best.

* * *

Fortunately, one of the great things about running, at least distance running, is that on the whole it is honest. Recall that when running a 10,000-meter or a marathon, you are operating a whole machine of an almost incalculable number of parts that are integrated, with no one part taking precedence over any of the others. You can't just increase one function and expect improved performance. Nature does not put in extra parts, nor does she give extra capacities to given parts. You have to improve *all* the systems at once to affect the whole. Can any drug do that? Possibly, but improving one part by a simple fix simply transfers the limiting factor to the next link in a long chain. The one factor that *does* affect the whole body, all at once and in a coherent, coordinated fashion, is in the mind, where courage comes from. I put more stock in placebos, faith, and specific training for a desired event.

Simply running to train for speed and endurance stresses *all* relevant points—*all* of the innumerable links in the very long and very complex chain—at once. That is, the necessary complexity and efficiency of running is achieved simply—by running. I did not need to lift weights. I never lifted a weight again in my life after I hurt myself, nor have I ever taken even a whiff of any performance-enhancing chemical. To become a runner, I just ran.

Coach Styrna, who had studied at the University of New Hampshire, told us how he had taken many courses from which he learned things he never used. What he learned in track and field he needed for everything in life. For example, you don't get anywhere by magic, but only by putting in the required number of steps, one at a time, and in the correct sequence. You can't run the last lap of a mile until you've run the first three. There is a truth, a beauty, and a symmetry in this that is inviolate. Every step counts. Each is an act of beauty. Together they create *stride,* and in terms of the whole, *pace.*

Studying physiology and running appealed to me. My pace as a university student and athlete both improved, and prospects for actually getting a degree became promising. During my second summer, I worked up in the Maine north woods of Aroostook County again, but this time I was not alone. Far from it. I was in an old-fashioned lumber camp populated by a couple of hundred French Canadian lumberjacks, who came across the border each summer to cut pulpwood for the paper mills. With several other forestry students employed by the International Paper Company for the summer, I walked through the woods for five days a week with a spray-paint can. We marked trees for selective cutting. (The company later abandoned selective cutting and reverted to clear-cutting.) We awoke to the cookie's (assistant cook's) clanging bell at dawn, jumped out of our bunks, raced to the cook shack, and took our places on long log benches in front of tables laden with eggs, bacon, cereal, doughnuts, cake, biscuits, coffee, tea . . . After our gargantuan breakfast we headed into the woods, coming back out at five in the afternoon for a lumberjack meal of meat, potatoes, vegetables, pies. I rested briefly on my bunk after the big supper, then got up again, and with my boots still on ran several miles up and down the dusty road from our camp. I believe it was this active regimen that ultimately healed my back, which had not been expected to heal.

Well nourished and in excellent shape when school started in the fall, I was even more motivated because of my wrenching setback the year before. This year I flew. I was with the leaders of the pack, and my spirits soared. I joked with my team members that we were animals. We mock-growled at the starting line before a race. The better we ran, the more we all felt that the word "animal" was an accolade. Indeed, we won not only the state and Yankee Conference titles; we also got the chance to run in the Eastern Division of the Intercollegiate Amateur Athletic Conference, held in New York City that year. And wonder of wonders, we won it. We were the best team east of the Mississippi in our division. My roommate and teammate, Fred Judkins, from Upton, near my hometown, won the overall individual honors, an incredible victory. The night before our race, we ate at a restaurant on Madison Avenue. He amazed us all by consuming two servings of steak, two huge baked potatoes, and *two* servings of apple pie and ice cream. Fred went for doubles in everything. Later this included two tours of duty in Vietnam as a helicopter pilot.

I settled into my studies and began to tolerate the work of washing dishes in the cafeteria kitchen and my other job, collecting dirty coffee cups in the student union. To my surprise, some of the subjects were beginning to interest me, and I even managed to earn B's instead of the usual C's. Indeed, I soon made the dean's list (B or better average). This was not unusual for the track athletes, but it was quite an achievement for me—I had received rejection letters from four of the five colleges I had applied to. Coach posted a write-up from the school newspaper showing that the members of our track and cross-country teams racked up the highest grade point averages of any groups on campus. David Parker, our best sprinter and state and Yankee Conference quarter-mile champion, even got straight A's, an incredible achievement for an engineering-physics major.

Athletic letters and awards were handed out at a special banquet in the winter. I'd earned my big blue *M,* the letter signifying I'd made the cross-country team. A dark blue jacket with a large, light blue letter *M* emblazoned on it was also part of the prize. Like all lettermen, I would wear it proudly. After the awards there was an election by team members to choose the cross-country captain for next year.

When the votes were cast after our lavish meal and all the speeches, I was not just curious, I was nervous. I was tempted to vote for myself. Who knows? I might break a tie. But I resisted. We wrote out choices on strips of paper, folded them, and brought them to Coach, who was sitting with all the other dignitaries at the head table. He unfolded them one by one and put them in separate piles. Then he made an announcement: "The captain chosen for next year is—Ben Heinrich." I was choked up with emotion.

The university was a fabulous place. I loved going to the difficult and exciting classes with friends, then unwinding in the Bear's Den between classes at small, intimate tables, drinking our coffee. We'd talk with the smiling, beautiful coeds, watch the fraternity brothers with their different bright-colored frat jackets. I was not one of the fraternity men, in part because I didn't have time to party and I had a large deficit of social skills. But the cross-country team had given me their ultimate social accolade. I could never let them down, especially as it was likely I'd really be needed, for we had the chance to be state, if not New England, champs again.

But I did let them down. Shortly after the team elected me to be their captain, Papa, then in his late sixties, announced that he and Mamusha were about to make their last great expedition—and he wanted me to go with them. It would be to Africa, the mysterious continent I had read about in the books of Osa Johnson, Carl E. Akeley, and others. Africa to me represented the ultimate in adventure. There was no question that this would be a once-in-a-lifetime experience, and it was my only opportunity to be with my parents and to see the life I had heard so much about when, as a child in the Hahnheide and during our first year in Maine, Papa had taken us two kids into bed with him to tell us about his adventures in far-off jungles. Having spent nearly six years at the Good Will orphanage after that, I had seen little of my parents since we had come to America. I wanted to give Coach and the team my very best efforts, but after painful deliberation, I knew that there was no choice: I had to go to Africa. Years later, I would realize how profoundly this trip influenced my thoughts about running.

My job during my thirteen months in Africa in 1961 and 1962 was to hunt birds and to skin and prepare them for a museum research collection. There was never a day off. For me it was a nonpaying job, because Papa felt the privilege to go along was compensation enough. He and Mamusha were employed to collect rare birds in isolated forest islands for Yale's Peabody Museum. Along with several local Africans, I was just their helper. My only diversion from bird hunting was to collect insects for my father's collection.

We lived and worked in tents the whole time. Mine was a pup tent with a bedroll. I lived in the field for most of the day with a shotgun in my hand and a bag on my back. At dusk, I sat around the campfire with Mohamed, Waziri, and Baccali, our African helpers, and shortly after dark I crawled into my pup tent to sometimes scribble in my diary by the light of a candle. At dawn, I crawled back out to eat the oatmeal Mamusha cooked and to return for the day, alone, into the forest. I was a predator pure and simple, and at age twenty-one, I loved it. There was no need for pay. I could spend days tracking down new voices, new birds, and never grow tired of it. I felt as I had in the Hahnheide, only now the smallest bird was often the most valued prey. Food was a distant, secondary objective. The hunt itself was its own reward.

In the desert scrub, with Mount Kilimanjaro in the distance, I wandered daily and heard spur fowl, guinea hens, hornbills, duetting barbets, and untold hundreds of other birds. I watched bateleur eagles and vultures soar overhead. I smelled the sweet scent of acacia blossoms, and I saw them abuzz with brilliant cetoniid beetles and butterflies. I walked under massive baobab trees with gray, wrinkled bark like elephant's skin, honeybees streaming from their cracks and folds. I saw the spoor of dik-dik and gerenuk. I saw a red, sandy path with human footprints leading to thatch-covered huts. In the evening, I smelled wood smoke and heard the throbbing of distant drums.

I had no reason to run, but when we were near a dirt road, I usually did run when our busy schedule allowed. On one occasion, on the slopes of Mount Meru, I ran barefoot as I had seen Africans do and as the Ethiopian Abebe Bikila had just done while winning the Olympic marathon in world record time. However, by my usual turnaround point at a small pond I saw blood between my toes, and I would have liked to stop. Instead, it was essential to speed up because it would soon be dark, and predators could ambush no matter how alert I was. Adrenaline dulled the pain and I got back to camp safely, but at a cost. The soles of my feet had become like raw hamburger. I lost the skin off the bottoms of my feet, and could not walk for about two weeks. In running, long physical training is essential for toughening everything from the head to the legs and all the way down to the bottoms of the toes.

After returning from Africa, my running at UMO was, in retrospect, good but not outstanding. I injured myself again, breaking some cartilage in my knee while pushing my hundred-dollar little red Hillman down a hill to get it started, as always. Eventually the knee required surgery. The Hillman was beyond repair.

I did not feel I was within close enough range to aspire to anything outstanding. Lettering in the track and cross-country teams, and sometimes even winning my events, was almost more than I had hoped for. Except for just possibly one thing.

The names of all the previous heroes who had set university records were inscribed on plaques hung on a wall along the field house cinder track. I'd run by them thousands of times—the names and achievements of the athletes of the past who were now legends. A few of the current crop of athletes would manage to post their names while I was still there. There just was no way, it seemed, that I could rank among them. Eventually, I improved in the 2-mile run, my specialty on the track team, and it seemed remotely possible that I might leave the university with my name on a plaque. Chasing a realistic goal injected my running with new fire. Finally, by the very last race that I'd ever run in the field house, I thought I could do it. My confidence seldom exceeded my ability. This was it. I'd go for broke.

There were too many laps per mile in the confines of the field house for me to keep count while I also concentrated on pace. My priorities were to concentrate on running smoothly and to focus on lap times, until I knew I was on record pace. Then I would just hold that pace and hope to hang on to it through the final, or gun, lap. At the sound of the gun signaling the beginning of the last lap I would simply fire off and let it all pour out.

All was going according to plan. I heard Coach yell out the times for the quarter, half, and the mile. My intended pace was right on, and I was feeling strong, saving up for the big kick. I could hear people yelling as they crowded to the edge of the track. The excitement mounted. I reached deep. One more lap. Another. Another. I soon anxiously waited for the gun lap, eager for the final sprint. More yelling. Then I heard the long-awaited *bang.* But strangely—it was not just one bang. There were two in rapid succession, and they had come while I was already at least 10 yards *beyond* where the lap should have begun. No matter. Maybe this was for emphasis. Maybe they knew I was on record pace. I reached deeper, found what I wanted, and took off.

That last lap was probably the fastest postrace lap ever; I had run an extra lap. The race official with the gun had lost track of the lap count himself, so he had given me not the signal to sprint in, but to get me to *stop* running. I had already missed the record, but only by a mere two-tenths of a second. When the actual final lap had come, I thought I still had several to go, and when I did pour it on, the race was already over.

That this mix-up would happen to me, and never to anyone else (to my knowledge) at UMO, and then on the very final of the very many races I had run—the only one where I assuredly would have set a record—seemed almost bizarre. But it was consistent with many other experiences before, in which fate had led from bad to good fortune. And so I wondered what good fortune might come from this teasing disappointment. Maybe it fanned the running flame or redirected it. There may be no more compelling goal than a close one, but there is none more lasting than a distant one not yet attained. The result of this race gave me a long-lingering sense of disappointment that contributed to my desire to go for it again, but at a much longer distance, some seventeen years later. For the time being, I continued to run for fun, but I was sure that my racing days were over.

Runners are rational. They have to be. Although driven by dreams, they learn uncompromisingly to confront facts, and they do not get misled too far by wishful thinking. They are like wolves and other predators who hunt by choosing their prey wisely. They do not chase fleet-footed antelope that they *know* will outrun them. Back in the 1960s and 1970s there was still much reachable "game" to catch by those who would put in the work and who had the dream. There were, therefore, many "hunters," so that a constant new crop of talent was exposed. Some of them, it turned out, possibly to their own surprise, were blessed with great talent. They became the heroes of my boyhood and beyond—Jim Ryun, Herb Elliott, Peter Snell, Frank Shorter, Bill Rodgers, Billy Mills, Lasse Viren, Steve Prefontaine . . . They aspired to be gods, and at some level they were. Yet the real reason my high school and college running mates and I saw them as heroes was that we secretly believed we were elementally equal. We were convinced that if we only tried, if we did what they did, then we too would rank among the gods.

Now all the records have been raised to unimaginable heights. The results are dramatic. Kevin Setnes, an ultrarunner and coach, wrote me in June 2000:

> I don't think there is any question that America has hit rock bottom when it comes to running performances. Some records have been set, but except for the few exceptions, runners today are way behind their predecessors fifteen years ago. Boston's [marathon] tenth American finisher today would have been hundredth in the mid-80's. In 1983 I was 51st American in the Duluth marathon, in 2:25. This past weekend an eleven minute slower time got tenth place. The Olympic Trials is not just another example in a long list.

There are now many more runners out there than there have ever been. But they run for the enjoyment of it, for health, and possibly for social reasons such as companionship. Who among them that can run a marathon in 2:20 (a time that would have left *all* the great marathon and distance runners in the dust half a century ago) would now seriously dream of, much less plan on, running to glory in the local Boston race, knowing that large contingents from other countries will stream by as if he were standing still, with many going well under 2:10 to claim the glory and the prizes? I know I would not. Back then, after my failed attempt at the field house 2-mile record, I knew that it was time to do something else.

At Good Will School, I had grown up reading the adventure stories of great explorers and scientists. There were many real heroes and fictional characters like Professor Gottlieb, idealized in Sinclair Lewis's *Arrowsmith,* who elevated test tubes and Bunsen burners to the level of sacred objects. Professor Gottlieb labored alone in the laboratory. I felt inspired by his image, because a scientist's achievements, like any athlete's, are produced by individual effort and by taking one little step at a time in a consistent direction. These men and women of science were heroes and role models. They did experiments to see what made animals tick. They did not engage in hype or genuflection to seek influence or beg for money.

After graduation and receipt of my 4F military draft status, I had worked for Professor Dick Cook washing glassware in his laboratory. Dick Cook to me was Professor Gottlieb incarnate. I learned from him how to flame test tubes just right with a Bunsen burner and how to raise and maintain sterile *Euglena* cultures. Before I knew it, I was studying *Euglena* cells' respiratory physiology and metabolic pathways. One day, Dick said, "Why don't you make this work your master's thesis?" He was proposing a new chase, one that would have seemed inconceivable to me just moments before. Suddenly it seemed rationally possible. This time I would not be striving for a record to hang on the wall in the university field house, but instead I might make discoveries. The lure of the hunt in this new game had been redirected, but ironically the research I did may indirectly still have related to running.

One of the main points of athletic dogma is that our rate of oxygen consumption equals our rate of sustained energy expenditure. Thus, our $\dot{V}_{O2}$ max is thought to be an accurate predictor of maximum work output, or ability to run fast for long distances. Often it is, but had Derek Clayton, Frank Shorter, or Alberto Salazar known it, they might not have even tried to set world and Olympic records, because their $\dot{V}_{O2}$ max is near 70, almost modest in comparison to, for example, Steve Prefontaine's 84.4; he could run a 3:54 mile, but his marathon performance does not even come close to that of the three others, who set world records, won Olympic gold, and were *the* kings of the road. The fact is that some people get more aerobic work out of the same volume of oxygen they inspire. That's what's called efficiency. Children are notoriously inefficient runners because they breathe in a lot of oxygen but (relative to trained adults) don't run very fast even when running at a high percentage of their $\dot{V}_{O2}$ max. I was probably lucky not to have been typecast for aerobic capacity or I might not have pursued anything requiring lung power, because it turns out that what you have is less important than what you do with it.

Running efficiency involves neuromuscular mechanical coordination that is probably acquired through years of training. You see it in a smooth, fluid stride where there is no wasted motion, no up-down bobbing, no lateral arm swings, and of course, no carrying of extra weight. Given the physiological studies I did with Dick Cook at the University of Maine, I could not help but wonder if metabolic efficiency at the cellular level might not be involved, as well as mechanical efficiency, in making the runner. We grew the single-celled protozoa *Euglena gracilis,* which move themselves around in the water by vigorously whipping a tail, or flagellum. They are metabolic virtuosos that can live off sunlight as an energy source. They can also get their energy from acetate, short, two-carbon fragments that our own cells use when we break down multicarbon chains of fats to use for energy. Finally, they can also get energy from common glucose, the sugar that we get when we break down glycogen from its storage form in the muscles and liver, to use for energy.

Dick and I fed *Euglena* by dissolving either acetate or glucose into the watery medium in which we grew them, and then we measured their rates of oxygen uptake. To our great surprise, we found a massive, four-times-greater rate of oxygen consumption (that is, presumed metabolic rate) in the acetate-using cells than in the glucose-using cells. Yet rates of cell protein synthesis, growth rates, and mass were identical. When we gave glucose-adapted cells acetate, they showed no immediate effect of increased respiration. However, after a period of adaptation, during which we measured sharp increase in the enzymes for acetate use, the glucose-grown cells began to use large volumes of oxygen. In both acetate- and glucose-grown cells that we then washed to measure oxygen uptake in the absence of substrate, the respiratory rates were equal. Our results showed that acetate-using cells have the same net energy production as glucose cells, despite consuming massively more oxygen. Additionally, they produced an acidity that soon killed them, whereas glucose-grown cells continued to divide and reproduce, staying healthy.

I'm not implying that we can directly apply these results to runners. However, they do show that oxygen consumption rate, the basis of $\dot{V}_{O2}$ max, is not necessarily totally diagnostic of the cells' sustained rate of energy expenditure. This alerted us to the possibility that metabolic efficiency could be as important as mechanical efficiency.

There are innumerable steps involved before an ingested food molecule finally has the energy it contains converted to adenosine triphosphate (ATP) and creatine phosphate (CP). The ATP and CP are the immediate energy sources used in muscular contraction. There is loss of potential energy at every step in a complicated conversion process, so the scope for cellular inefficiency is large if the number of metabolic steps are many. Such cellular inefficiency has not been considered as a potentially major variable in athletic performance. If it is a factor, then we can predict that there would be a lot more *variability* in endurance versus sprint or weight athletes, simply because in the endurance athletes' reliance on mitochondrial respiration, there are innumerably more metabolic steps contributing to and directing throughout the whole performance. A sprint athlete can run 3–5 seconds simply off the ATP that he has made previously from anaerobic metabolism occurring in the cells' cytoplasm, not within the mitochondria. Further anaerobic metabolism from the breakdown of glycogen gets him up to a half minute or so farther. During a long run, the distance runner has to continually operate every single step of the energy-yielding operation, all the way from the stomach (at greater-than-marathon distances) to the mitochondria within the cells, to produce ATP for muscle contraction from aerobic metabolism.

Looking at the metabolic energy production of a microorganism such as *Euglena* may not be entirely frivolous, because mitochondria, the sole organelles of our bodies that supply us with *all* of our aerobic power and give us the ability for *sustained* exercise, are evolutionarily derived from bacteria. It seems rather arbitrary whether we call them "cell organelles" or "highly adapted bacteria" that use us as carriers to house and propagate themselves. Mitochondria still contain their own DNA, which is much more variable than our chromosomal DNA. It can, therefore, be suspected that their metabolism, which produces the ATP we use for muscle contraction, is also variable. Hence, the metabolic efficiency of aerobic metabolism could differ among individuals, affecting either their $\dot{V}_{O2}$ max or the power output per given $\dot{V}_{O2}$ max. Each of us is seeded with mitochondria strictly from the female line, through the egg. This implies that, if mitochondrial efficiency is indeed a variable in aerobic capacity and we want to be champion *distance* runners, then we must look closely to the *maternal* line if we want to choose our parents wisely.

In two years, I'd published three scientific papers with Dick Cook on *Euglena* metabolism. I was hooked, thinking I was leaving a mark that would be more permanent than a name written on the field house wall. We had discovered a new metabolic pathway, and I was never pushed to the edge of my capabilities as I had been routinely in track, in cross-country, and in my undergraduate studies. We showed up at the lab every day, and it was so exciting to me that I couldn't even stay away on weekends.

I gave an oral presentation of my work to the whole assembled faculty and zoology graduate students, and as we walked out of the amphitheater together afterward, Dick pulled the pipe out of his mouth and said in his soft, low voice, "Ben, that's the best damned seminar I've heard here for a long time." That was obviously a damn lie in the sense that others, more interested in something else, surely thought it trivial or incomprehensible gibberish. But that was fine by me, as long as Dick was pleased. Later Dick said, "You've actually got nearly a Ph.D. But I won't give it to you. You need to leave Maine, to get more experience."

As I've said, *Euglena* can become totally independent of energy from *food*. When given light they quickly develop their own populations of energy powerhouses—little green organelles, called chloroplasts, that harness the energy from the sun directly. Like the mitochondria derived from bacteria, chloroplasts also have an ancient origin; hundreds of millions of years ago they were derived from free-living algae that invaded *Euglena* ancestors, and some of these associations became plants. They, like the mitochondria in other hosts that become animals, also still have some of their own DNA.

Dick had got me started assaying RNA and DNA from *Euglena* during the cell-division cycle, but we had no way of differentiating the potentially different kinds of DNA. During one of our discussions on these matters during a long car ride on a fishing trip to the Narraguagus River, Dick suggested that I ought to go to UCLA for my Ph.D. "They have a great bunch of protozoologists and molecular biologists there. You should study extrachromosomal DNA. We know hardly anything about it." Nothing could have sounded more exciting to my ears.

Having money left over from my research assistantship, I bought my first real—that is, functionally starting—car, a secondhand white Plymouth Comet. Packing nothing more than a box of clothes and a sleeping bag onto the backseat, I drove across the country, pulling over onto some deserted road at night to sleep and having breakfast the next morning at the next available diner. After I crossed the state line to California, I went straight on to Malibu Beach, to try surfing. In the afternoon I drove through the redwoods, and that same day walked the halls of the zoology department on the UCLA campus, where I met a student whose neighbor had just moved out and who needed someone to take over the apartment. Within a week, I met Kitty Panzarella, a fellow student who soon moved in with me into that little apartment on Greenfield Avenue, and who became my wife.

In 1966 Los Angeles was quite an experience for someone straight out of the Maine woods. I recall scenes of driving down the Santa Monica Freeway with six lanes of traffic all in one direction, and as many in the other. A blue haze overlay endless tracts of flat houses surrounded by palm trees, and through the distant haze I saw a gigantic labyrinth of metal pipes, tanks, and chimneys spewing white fumes. As we were cruising down the freeway in this landscape, the driver of the car (a fellow graduate student) had the window open, and over the radio, turned on full-blast, I heard Jim Morrison of the Doors pounding out a hypnotic rhythm: "Light my fire . . . light my fire . . . try to set the world on fire." Not very likely. We were sweating, sipping beers, and heading to check out a love-in at Griffith Park, where crowds of stoned hippies with long hair and beards, dressed in bell-bottom trousers, were smiling beatifically. It was surreal, unearthly. I could just as well have landed on the moon.

Despite a supportive mentor, a loving wife, and a generous research assistantship, I still had not made much, if any, progress on a thesis project, even after one year of hard trying. The DNA trail was growing cold on me. I felt like a miler who suddenly realizes he doesn't have natural speed, and so can't realistically expect a great prize, no matter how much work he puts in. I was wandering and wondering more and more, with fewer rather than more prospects of catching something big.

There was only one thing that stayed familiar. The tartan track. I was on it almost every afternoon or evening, just to run. I had no thought of racing, although a group of us regulars became an informal club, and we called ourselves the Termites. Termites are paragon social animals. They are not very swift, but by sheer dint of persistence and teamwork they chew through solid wood. We ran mostly quarter- and half-mile intervals, in little Termite bands, and in the UCLA intramural competitions I got roped into running the mile. One of the guys on the team talked about someday racing 26.2 miles—a marathon. I thought he was nuts. I could at least potentially wrap my mind around mitochondrial and chloroplast DNA (although doing so "with my fingers" eluded me), but I could not comprehend the possibility of jogging, much less racing, over such a long distance.

SEVEN

How to Reduce an Insect's Flight Endurance

Ask now the beasts, and they shall teach thee.

—JOB 12:7

Hunting for scientific discoveries is a game. It's much like hunting for wild game. Proximally, both are done for the fun of it, and both ultimately yield practical results. The problem is that life's thickets are incredibly dense, and you have little idea what is lurking in them. You may want to go after a big payoff, if one is there and you're sufficiently gifted to identify and pursue it. But there is no guarantee you'll find it. I presumed, to get a Ph.D., I'd have to discover something totally unthought of and amazing, perhaps equivalent to discovering DNA, deciphering the genetic code, or elucidating the control of mitochondrial growth. As I already mentioned, my efforts were frustrating. Furthermore, as at UMO, my own physical ability to move soon fell short as well. After being unable to get data from DNA and realizing I wasn't cut out for spending the rest of my life at a molecular biology workbench, and needing a new direction, I developed weird arthritis pains in the hands, knees, and feet that put me on crutches for a half year. As at Maine, the physical incapacity gave more time to spend in the library; then I made research probes to examine various aspects of behavior and physiology in tiger beetles, honeybees, caterpillars, butterflies, and hawk moths, trying to discover something of sufficient fun, that is, something new and of intellectual value. I eventually focused on problems in exercise physiology and temperature regulation in hawk moths. This choice was fortuitous, because it led to much more than I could have imagined.

What might insects teach us? Insects are creatures so different from us that they could have evolved on another planet. They have no brain as we know it, but instead a series of variously sized clusters of neurons. They have no veins, no liver, no bones, no lungs, no kidneys, and a very much different set of hormones. Except for those desert cicada active in summer at midday, insects don't sweat to dissipate heat. Their "skeleton" is on the outside, rather than inside. They have no hemoglobin because they do not use their blood to transport oxygen as we do. Instead, small tubes called tracheae lead directly from closable portholes along the outside into the cells, with no intervening circulatory system. Nevertheless, despite our huge physiological differences, they solve similar problems as we do and by a number of accounts they are the most successful animals of the planet.

I knew hawk (or sphinx) moths and their larvae fairly well, and George Bartholomew and Franz Engelmann, my

Hawk moth while feeding

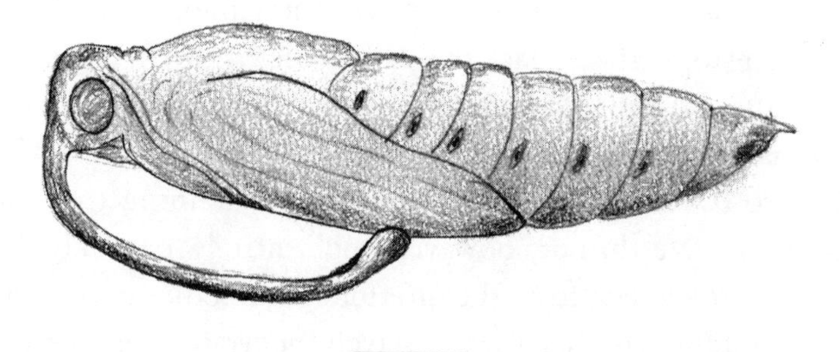

Hawk moth pupa

thesis advisers at UCLA, alerted me to a previous publication that suggested that these large insects might regulate their body temperature in flight—that is, keep it stable regardless of differences in external temperature. Since they fly at night and cannot bask like lizards and butterflies do, it seemed to everyone that they did this trick metabolically. Nobody had a clue as to how they might do it or if they really did. The topic concerned exercise, and my data from the moths soon related to their flight endurance.

Unlike butterflies and bees, foraging hawk moths are continuous fliers, which, like hummingbirds, hover and fly from flower to flower without landing on them. They are hot-bodied only just before and during flight. Unlike a hummingbird, after perching and coming to rest, a moth's heat production stops instantaneously, and body temperature cools by passive convection to become essentially identical with air temperature within a minute or two.

Heat loss by convection into air is perhaps most easily explained with an analogy. Heat in the body convectively lost to the air is like dye leaking out of a cloth bag into water. The rate at which the dye diffuses into the water depends on the color gradient at the bag-water interface. Finally, when the color inside the bag equals that outside (that is, when the temperatures are the same), then all convection stops. If we place the permeable bag with dye (heat) into a fast-flowing stream (wind), the rate of dye loss (cooling) is greatly accelerated because the dye immediately adjacent to the bag is quickly removed, maintaining the color gradient. We do not, however, cool entirely passively. We actively *shunt* heat from the interior of our bodies to the skin, to maintain a higher than passively generated temperature gradient. We also sweat, which permits us to lose heat *against* a temperature gradient, that is, from a lower internal to a higher external temperature.

Like us, the moths need a high muscle temperature in order to exercise vigorously. When their environment is cool, they achieve that muscle temperature by a slowly accelerating exercise—by using their flight muscles for shivering prior to flight exercise. They do not, however, produce more heat in flight at lower than at higher air temperatures. Similarly, when we run at low air temperatures, we don't use our leg muscles simultaneously to shiver. By running, we already increase our body metabolism from 1.5 kilocalories per minute up to 30, but we can't shut the heat production off. It is an unavoidable by-product of heavy exercise even if we're running on a hot day. Heat production and exercise are inexorably linked. We can reduce heat production only by slowing down, but this option is limited for a hawk moth, which needs to expend huge amounts of energy hovering in order to eat. In summary, heat *production* in the moths turned out to be strictly a by-product of energy expended for flying. Yet paradoxically, body temperature during flight remained stable within a remarkably wide range of air temperatures over which passive heat loss would vary widely. How could they keep their body temperature both high and stable in the face of a wide range of environmental temperatures where the rate of passive convection would be expected to vary hugely?

Not being able to reduce heat production when we begin to overheat while running, we instead sweat to get rid of extra heat. We can thereby keep on running and producing even more heat without overheating as long as we have enough fluids to sweat. I found no sweating in the moths, yet they still stabilized their body temperature. How they kept warm at low temperatures was clear, but how did they cool themselves at high temperatures? How do they get rid of extra heat to stabilize their muscle temperature in order to continue flight at high temperatures? To find out, I performed a series of experiments that proved they get rid of excess heat from the muscles in the thorax by a unique mechanism. They shunt the heat from the thorax into the normally cool abdomen using the blood as a heat-transfer vehicle. The abdomen has little insulation, so wind passing over it causes cooling by convection. This so-called convective cooling from the abdomen is comparable to the heat dissipated from our car radiator, after it is transferred from the engine by liquid coolant.

The moth's abdominal heat radiator could keep the animal from overheating during continuous flight even at air temperatures of 30°C (86°F). However, I reduced my moths' flight endurance to a mere two to three minutes by surgically doing the equivalent of crimping a car's radiator hose—by tying off their heartlike structure that pumps blood. The operation destroyed the moths' ability to transfer heat into the abdomen; thorax temperature shot up, while the abdomen stayed cool. The temperature of the muscles in the thorax that power the wings in altered moths rose explosively to the intolerable high temperature of 44–45°C (111–113°F), and like marathoners who run out of water for evaporative cooling, these animals dropped to the ground with heat prostration. I knew that the overheating, and not the incapacitation of their blood-pumping organ, was responsible for the moths' limited flight endurance, because if the fur coat covering their flight engine, the thorax, was removed so that they could passively lose more heat, the altered moths flew well despite the heart operation. It may seem counterintuitive that the moths have a furry thorax, which helps keep in heat. Thick fur is indeed a liability during flight at high air temperatures, but it is useful at low temperatures, when the animals have the opposite problem.

I was surprised and pleased with these and related results and published five different papers, three of them in the prestigious journal *Science*. There soon came a flurry of other papers showing that exercise endurance in various mammals is also limited by overheating. Jackrabbits, red kangaroos, and cheetahs are all furred to keep warm, yet all were shown to overheat to the point that they had to stop running even in moderate heat. Humans, however, because of a superb sweating response, have remarkable running endurance in the heat.

The problem of sometimes having to stay hot to exercise,

and at other times having to get rid of heat to continue to exercise, can also involve precisely synchronizing the breathing cycle with blood circulation. As I shall next explain, breathing causes blood to be pumped and heat to be dumped, and that gives some insects endurance in the heat.

To begin to establish the elegance of the insects' solution, as discovered in bumblebees, we have to back up and review some basics of design. First, in bees the abdomen is attached to the thorax by only a tiny, narrow waist, the petiole. All of the heat that in flying bees is some hundreds of times above resting metabolism (the exact number depends on what body temperature is used to establish resting metabolism, to make comparisons) is generated by the flight muscles that pack the thorax. (In insects there are no muscles *in* the wings—all the muscles that power the wings are inside the body itself.) The abdomen's temperature is generally close to air temperature, unless, as in moths, the abdomen is used as a heat radiator to get rid of excess heat from the thorax.

The flight muscles are totally aerobic; like distance runners, bees don't go into anaerobic debt as sprinters do. Their large $\dot{V}_{O2}$ max is made possible with the aid of air sacs in the abdomen. The whole abdomen pumps in and out as a piston, compressing and expanding the air sacs by positive and negative pressure. Those same pressure changes that pump air in and out of the thorax are also harnessed to help pump blood, and that blood either may or may not be used for heat transfer. When the blood is used to get rid of excess heat from the thoracic muscles, then the ventral diaphragm (see page 99) releases the hot blood into the abdomen in discrete pulses, while the cooler blood entering the thorax from the abdomen is also "chopped" into discrete pulses. The hot and cool pulses of blood bypass each other because they are shunted through the petiole alternately, in synchrony with the in-and-out abdominal breathing movements. I therefore dubbed this process "alternating" fluid flow to distinguish it from blood flow going simultaneously in opposite directions in adjacent vessels, called "countercurrent" flow.

When the bees are flying at low air temperatures or discontinuously (such as when working on flowers), they have the opposite problem, of needing to conserve heat in the thorax. Under this situation, blood flow between thorax and abdomen is much reduced, and breathing movements are no longer of importance for moving the blood. Instead, the heart fibrillates and passes a slow, thin stream of blood into the thorax. This mechanism allows recovery of heat back into the thorax that would otherwise be lost to the abdomen due to the countercurrent flow (see page 99).

Physiological synchronization of different systems is now coming to be recognized for energy economy as well as temperature regulation. In running quadrupeds, especially in such energy-efficient distance runners such as dogs, there is a coupling of breathing with stride. The animal passively inhales as its front legs stretch forward, and it exhales as it pulls those legs back and the volume of the chest cavity is reduced. As a result, energy necessarily expended for striding reduces the amount of energy otherwise required additionally for the mechanics of breathing. Birds and many insects similarly harness the thoracic volume changes that automatically result from their limbs' movement, to help pump air. We humans, in contrast, were thought not to have such coupling of breathing with locomotion, so that we must invest energy specifically to expand and contract our chest to breathe. However, I know that in myself there is a very distinct coupling of arm swings with breathing. It's automatic, and hard to change. The coupling cannot save as

Bumblebee Body Temperature Regulation by Blood Flow

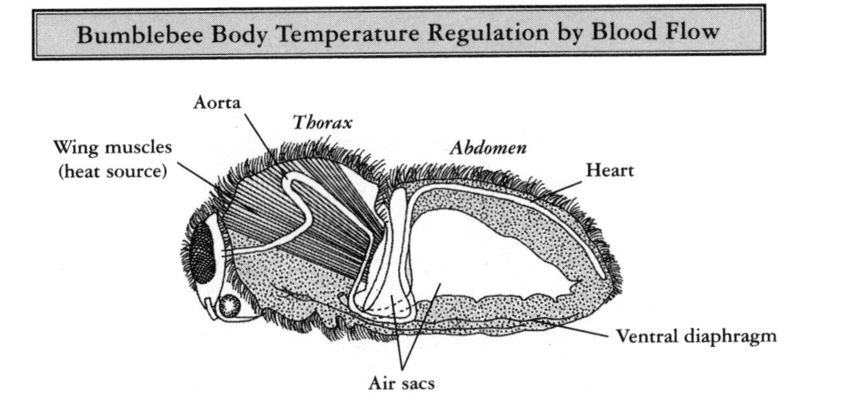

Heat Transfer to Abdomen (One Cycle of Alternating Current)

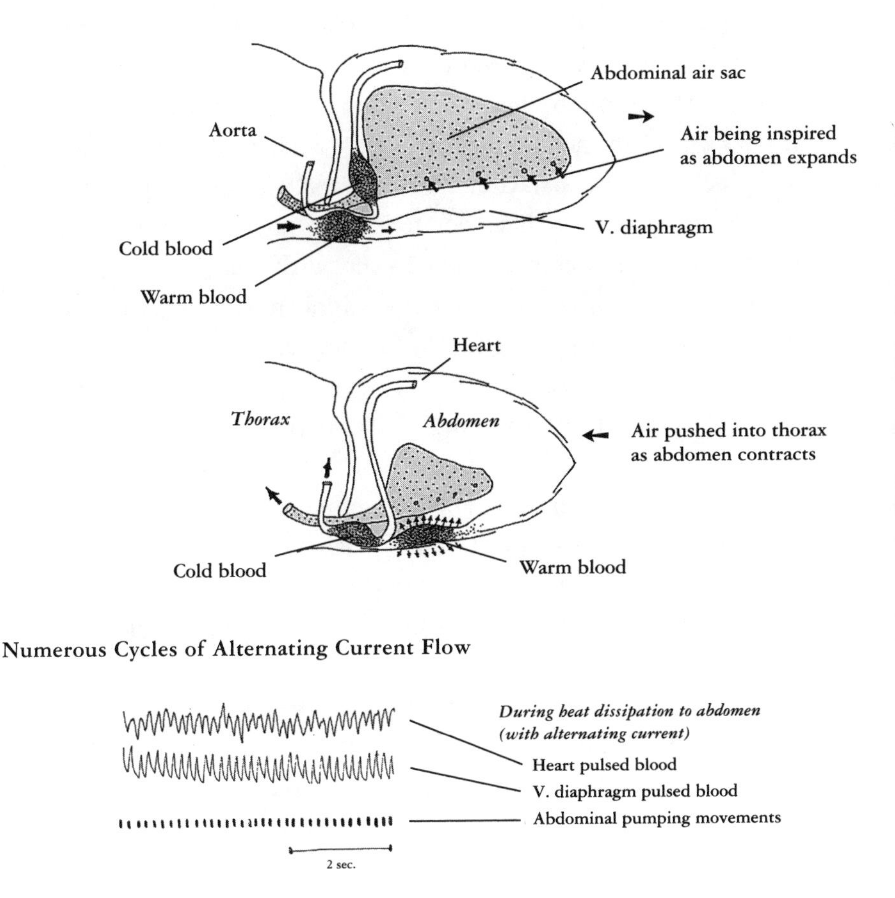

much energy as it does in a flying bird or a running dog, or as much as the well-known bellow-type breathing of some insects in flight does, but it surely saves some energy on a long run.

What applies to moths, bees, and dogs appears to apply to exercise in many other animals as well, and I became increasingly aware of my own breathing, heart rate, sweating, energy stores, stride, and running pace. When I ran, I sometimes tried to visualize as many of these variables as possible, trying to "see" how they all worked together. When I became conscious of it, I realized that at normal ultrarunning cruising speed I maintained a very specific pattern of breathing to my steps. There were almost always three steps with each breathing cycle, two for the inspiration and one for the exhalation. When I took longer steps, the only difference was that the inhalation lasted longer. The synchronization was maintained even with changing effort. When I was striding very easily, it took me three steps to inhale, and when I was running up a steep hill, two. In all cases expiration took place in only one step, and inspiration coincided with the remaining steps. When not running, I could also monitor heartbeats, and one heartbeat corresponded with inspiration and the remaining heartbeats of the breathing cycle corresponded with expiration. I'm not sure what the significance of that rhythmicity is, but I suspect it has something to do with energy economy. According to David Costill, ultrarunners have 5 to 10 percent less energy expenditure than middle and sprint runners per given running speed, and such running efficiency may take years of training to develop; it costs a twelve-year-old runner 40 percent more energy to keep pace with a twenty-year-old.

Synchronicity felt right to me, and so I probably unconsciously fell into the rhythm, and at times I tried to become conscious of it to facilitate it. By almost explosively exhaling on the last step of the breathing-step cycle, I maximized the amount of time the lungs held fresh air. Like a pronghorn antelope, I keep my mouth open when I run. Undoubtedly this reduces the resistance of expelling air and thereby saves energy.

One can predict generalities of evolution, given common sense. But you can't predict a bee's mechanism of temperature regulation. The common honeybee, for example, turned out to be an extraordinary endurance athlete in hot, dry, desert conditions. It could fly, even at air temperatures near 40°C (104°F), close to its preferred body temperature. Convective heat loss works only if there is a steep gradient of temperature between body and air. Without such a gradient, it is possible to lose heat only by evaporative cooling. So, given the first datum, the second is generally a necessary implication. However, bees don't sweat. This poses a quandary: What makes them special?

It turned out that they have a means of heat loss that is similar to one I saw used by rapidly overheating contestants on a race in over 80°F heat on the Bowdoin College track in Maine. Every couple of laps, the racers dunked their heads into a barrel of water the race director, Bill Gayton, had set thoughtfully alongside the track. The water evaporating from the contestants' heads and backs kept them cooled and running despite the heat. Surprisingly, bees that have collected nectar have a variation of this approach. They regurgitate their stomach contents from the mouth and spread the liquid all over themselves with their forefeet. Once they are back in the hive, colony mates lick off the residual solids (sugar) that are left after the water has evaporated. However, relying on regurgitation for evaporative cooling is probably not a recommended option for us.

Some storks and vultures cool themselves by a reverse, yet similar, strategy. They defecate runny feces down their legs. The blood in the bird's legs is cooled by the evaporation, which reduces overall body temperature by as much as 2°C. A turkey vulture sitting on a fence post in the sun on a hot day, calmly and deliberately defecating on its naked legs, is behaving in a way that makes sense. Anyone who has ever been running hard on a sweltering day will be able to identify with such behavior.

During my research with bees (later, at the University of California at Berkeley), I had corresponded with the world's premier biologist of social insects, Edward O. Wilson at Harvard University, and I was pleasantly surprised to learn that he once also had the runner's bug. Ed had aspired and now he inspired. Reviewing some of my running history, Ed declared out of the blue, "You could run a sub-2:30 marathon," and I immediately wanted to show him right.

Committing to run a marathon was a snap decision. I read his letter, and bang—I knew I could not allow myself one possibility of an excuse or of second thoughts. I could not say, "I'll start tomorrow." So I jogged to the gym, changed up, and ran Strawberry Canyon. I had visions of finishing in the Boston Marathon and then walking up to Ed's office at the Museum of Comparative Zoology seeing his broad smile as I'd scored one for *our* team, the one of biologists with butterfly nets and insect-killing jars.

Soon after I started training, I got a knee pain. I went to an orthopedic surgeon, who said, "You have (some sort of) cartilage degeneration. If you don't stop running, I'm going to have to take your kneecap off and throw it in the garbage can." His exact words. They rang in my ears a long time. I figured, instead, that I had a loose piece of cartilage, which I could get rid of by grinding it down by running, so I increased my mileage.

Ed turned out to be right, the orthopedist wrong. A half year later I did almost what Ed predicted. But I didn't bring him news. When I trotted up to his office, he'd already read the Boston paper, and he greeted me by calling out my finishing time: "Two twenty-five!" I'd run 5 minutes faster than he had predicted. This was something for both of us to enjoy. As is usually the case in science, you make a prediction, and if it comes out close, you are happy because you're potentially right with one idea, and if it comes out different, you're closer to some other idea that you didn't even think of before. That's even better.

During my training for the marathon, I had thought a lot about the exercise physiology of insects, especially that of my current subjects, the bumblebees, which I calculated on the basis of energetics to have a flight range roughly near my maximum running range, a marathon. But birds do much better.

Birds are a lot like us. They have roughly the same body plan, with the same organ systems. Like us, they have lungs, blood, true hearts, arteries and veins, liver, brain, and kidneys. They have the same basic types of limbs, sense organs, glands, hormones, and biochemistry. Their mechanisms of gas transport, immunity, development, excretion of wastes, and brain function are practically identical to ours. We differ from them mainly in how far and in what direction each of these common features has been stretched to serve specific adaptive scenarios. Birds are more highly evolved than we are regarding endurance physiology. They have achieved a true breakthrough in speed and endurance. Insects had revealed many secrets of endurance that relate to body temperature, but birds inform us about the ultimate endurance that is possible for flesh and blood.

EIGHT

Ultramarathoners in the Sky

Fall and spring, untold billions of birds take to the skies for nonstop flights of thousands of miles that often take them over oceans and deserts. Their very survival depends on their athleticism, mental resolve, and navigational skills. My race on October 4 in Chicago would be puny relative to what they accomplish routinely. I'd race for only 100 kilometers, with the route mapped out in convenient loops and with all the food and drink I wanted provided along the way. Like other birders, I have sometimes trained my binoculars up at night, seeing the migrants' silhouettes slip across the milky background of the full moon. I've heard the birds' faint cheeps in the dark, and I've wondered how our brethren, of flesh and blood like ourselves, accomplish their amazing feats. Could they teach us something about endurance?

I want to discuss how they survive migration, but we

Blackpoll warbler

first need to get a better fix on what they actually do. Answers are only recently coming to light, primarily as a result of the work of thousands of people all over the globe who have banded birds. An amazing picture is emerging that may seem scarcely more plausible than the old presumption that swallows must "obviously" overwinter in the mud rather than commute between continents.

I'm especially impressed by the small, beautiful, and exquisitely delicate wood warblers. Every June, the forests across northeastern North America resound with the sibilant lisping and chattering of the thirty-five species of warblers that have returned to set up territories, build nests, and rear their young. The journeying of one of these species, the blackpoll warbler, *Dendroica striata,* is now known better than most of the others and can serve as an example, although there are differences between species' travel routes.

By mid-July, the blackpolls' nesting in thick northern spruce or fir forests is complete, and both adults and young molt their feathers. The adults change from their bright nuptial garb to a drab and undistinguishable one. The molt is done in a month and the whole population, distributed from Maine to Alaska, begins to move. All the birds converge in the northeastern United States. Individual birds from Alaska and the west return to specific sites where they had been after previous transcontinental trips of thousands of miles.

After reaching their eastern seaboard staging areas, the birds, whose lean weight is 9 to 11 grams, now enter a phase of gluttony, technically known by the more polite term "hyperphagia." Taking advantage of the berries ripening at that time, as well as outbreaks of aphids and other insects, they double their body weight in as little as ten days. Most of this weight gain is fat that is stored in thick masses under the skin on the abdomen and on the chest just below the neck. Fueled up, the birds next converge on their final staging areas on Cape Cod, Massachusetts. From there, these tiny wraiths of the coniferous forest embark en masse on an awesome nonstop transatlantic flight of about 2,200 miles, all the way to Venezuela.

The starting gun that launches them on this nonstop ultra of ultramarathons out onto and across the Atlantic Ocean is the passage of a cold front. Flying at a little over 20 miles per hour, they are at first wind-assisted, provided they pick a cold front that results in southeasterlies. Gradually, the departing birds merge into flocks of about five hundred to a thousand individuals. By the second day, these flocks reach the still air over the Sargasso Sea, and after the third consecutive day and night of continuous flight they are boosted by the trade winds, and the now much leaner birds begin to appear on the northern coast of South America.

Blackpoll warblers are perhaps not so exceptional among bird migrants, but like the other songbirds that fill our sum-

Migrating sandpipers

mer woodlands, they are a constant reminder of avian competence. Numerous species of Arctic sandpipers breed even farther north and overwinter even farther south, thus engaging in even more stupendous travels.

The white-rumped sandpiper, *Calidris fuscicollis,* is one. This shorebird, barely larger than a sparrow, breeds north of the Arctic Circle. In the fall, like the blackpoll warbler, it migrates east across the American continent to the northeastern shores. Slimmed down, it again fattens up before embarking on a nonstop journey of 2,900 miles lasting at least three days and nights. At the end of that flight, the flocks reach Suriname, on the north coast of South America. As in all long-distance travel, energy supplies are vital for success. The birds require rich feeding areas for fueling up to be able to embark on the third and final leg of their long journey, this one of 2,200 miles overland across the South American continent, the Amazon, and on to Argentina, at the southern tip of South America, to complete a total trip of more than 9,000 miles that spans the globe nearly pole to pole. The birds' traveling involves a refined itinerary with major refueling stops at quite specific and essential wetlands and undisturbed coastline feeding areas. At the end of each of their epic fall and spring migrations, the birds again reach continuous daylight, after having just come from the midnight sun in either the Northern or Southern Hemisphere. In short, except for the days in transit, the world they experience most of the year is without nights; this enables continual feeding, frolicking, and flying.

Populations of the knot, another Arctic shorebird, have given us additional information, especially as regards migration energetics. Knots breed all around the North Pole, and they have widely divergent wintering areas. In the New World, the red knot, *Calidris calidris rufa,* makes a southward migration in the fall that covers approximately 7,800 miles. When not carrying fat reserves for migration, the birds normally weigh 120 grams, but when fully fat—just before takeoff—they weigh about 180–200 grams, and occasionally even reach 250 grams. Like the smaller songbird migrants, they regularly double their body weight prior to migrating.

When loaded to a body weight of 250 grams (130 grams fat), a knot has enough fuel for a theoretical maximum nonstop flight range of about 4,700 miles. Flying at a speed of 47 miles per hour, it has about 100 hours of flying time before it must stop to refuel. Knots make their migration in long nonstop flights between staging, or feeding, areas, where the distance covered in each flight is limited by the amount of fat they carry. Hence, the food supply at the feeding areas, where the birds forage for a week or two before resuming their migration to the next fueling stop, is critical.

Red knots leaving James Bay in northern Canada have three staging areas, or fueling stops, before reaching Tierra del Fuego, their ultimate destination, at the southern tip of

South America. Knots from the western Arctic first fly to the eastern American seaboard, refuel for two weeks, then fly on to Suriname, in northern South America, refuel again, then fly across the Amazon basin to southern Brazil, before taking off on their last flight, to the tip of South America. Here they are in continuous daylight while their breeding grounds are in the continuous night of Arctic winter. Like many other sandpipers, plovers, and terns, they span the globe from pole to pole.

Tremendous amounts of fuel are required for the birds' long flights, but weight is a burden as well. The sandpipers, like airliners, often fly higher than fifteen thousand feet, where the air is thinner and there is less aerodynamic drag. One downside of flying so high is that there is less oxygen available. To stay aloft in thinner air requires more speed to generate sufficient lift, which in turn requires greater energy expenditure and more oxygen to sustain the greater flight effort. This is a catch-22 situation, which the birds resolve by their digestive and respiratory physiology.

How have the birds' remarkable capacities evolved from their terrestrial dinosaur ancestors? When the ancestors of birds first took to air, there must have been strong selection for weight reduction. Bones became lighter, in part by becoming hollow. Further weight reduction was probably also achieved by changes in diet. To extract the scant energy resources from otherwise plentiful foliage, herbivores require huge stomachs and long intestines. No airliner can fly using plant fiber as power source; it needs highly refined jet fuel, with a high ratio of energy to weight, that can be combusted easily and quickly. To support the high-energy habit of flight, the ancestors of birds must have become selective in their diet, choosing fruit and insect protein over foliage. A high-octane diet of insects and fruit could allow jettisoning of an enormous volume and weight of gut, as well as of teeth and heavy jawbones for anchoring the teeth. Becoming selective in diet eventually would have allowed them to become even more selective, because they could travel farther. I suspect that such a self-reinforcing cycle may have triggered the virtual explosion of bird evolution that allowed this group of animals to become one of the most diverse, numerous, and amazing life forms on earth.

Even now, changes in the diet of some birds and in such omnivorous animals as ourselves proximally affect gut mass. Simply eating more protein results in a physiological response of reductions in gut length and mass. Although change in diet from low- to high-energy foods would itself have increased the power/mass ratio to facilitate long-distance flight, the real breakthrough in ultraflight endurance in birds may have come later, as a secondary consequence of having the body cavity less crammed with gut.

A body cavity that has less digestive machinery crammed into it has more space for other components. That space could either be filled with other organs or left as air space to maintain lightness. As it happened in birds, it was the latter that occurred, and ironically that's part of the great breakthrough that allowed them to fly fast even at high altitudes in an oxygen-poor environment.

Air-breathing fish, reptiles, and mammals are saddled with an inefficient in-out breathing system. We inhale into a baglike lung by raising our ribs and lowering our diaphragm to create negative pressure in the lung. We then push the air back out before we can take in another breath. It is practically impossible to deflate or empty our lungs totally since we can't totally collapse our chest. There is always some residual air left inside—air that becomes partially depleted of oxygen. When we inspire, we mix the new, fresh air that is fully saturated with oxygen with this oxygen-depleted residual air. Not so with birds.

Somewhere, sometime, in protobirds there occurred the great innovation of connecting their extra body air spaces to the lungs. This connection then made possible the routing of air *through* the lungs. With the aid of inflatable air sacs in the body cavity, birds now route air through their relatively rigid lungs, and they do it during both the inhalation and the exhalation phases of the breathing cycle.

It might appear at first glance that if air goes *through* the lungs, rather than in and out, birds don't exhale. However, they do exhale! The trick is that it takes *two* breaths for them to move a given mass of air in and back to the outside, as two separate slugs of air move through the respiratory system at the same time. The lungs are connected to a set of air sacs in front of and another in back of the lungs. Air goes through the lungs in both exhalation and inspiration because during the inspiration, when both sets of air sacs become inflated, the posterior air sacs become inflated with fresh air while used (stale) air (of a previous inhalation) passes from the lungs and into the anterior air sacs. During exhalation from the mouth, the air leaves the anterior air sacs to the outside, while the fresh air (from the previous breath into the posterior air sacs) goes into the lungs (see page 113).

These breathing innovations set the stage for birds to make further stepwise modification in almost all aspects of their existing morphology, physiology, and behavior. It has made them the most impressive ultraendurance machines of flesh and blood the world has ever seen.

Birds' unparalleled aerobic and respiratory capacity allows them to expend the necessarily high metabolic costs of flight even in the thin atmosphere above Mount Everest, where humans can barely crawl, as our $\dot{V}_{O2}$ max is scarcely

Diagrammatic representation of bird breathing. Mammals inspire by expanding the chest cavity, which is accomplished by contracting the rib muscles and the diaphragm. Air pressure in the chest falls, and air rushes into the lungs. Birds have air sacs attached to the lungs, but their lungs are relatively rigid; the air sacs expand and contract to ventilate the lungs. Both anterior and posterior air sacs expand during inspiration, the anterior with already-used air from the lungs, the posterior with fresh air. During expiration, the air from the anterior air sacs leaves the body, while the air from the posterior air sacs (from the previous inspiration) enters the lungs. The two-cycle respiration results in one-way traffic of air through the lungs.

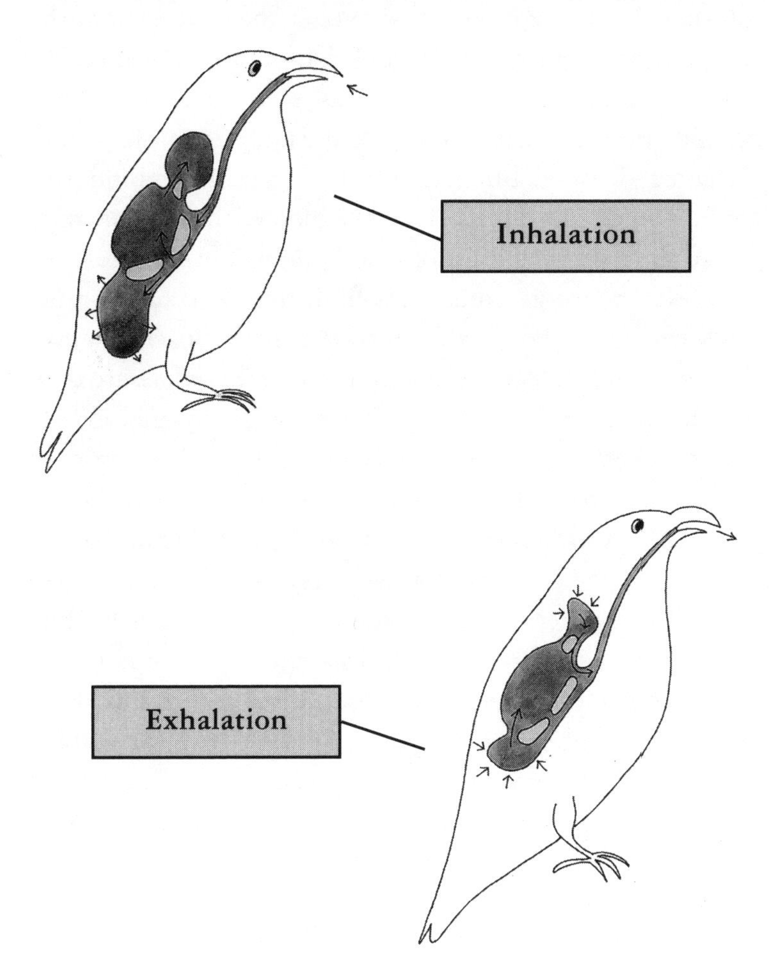

above resting levels. Bar-headed geese (*Anser indicus*) fly over the summit of Mount Everest (8,848 meters), making a journey of about 900 miles nonstop. How do they and many other high-altitude migrants get sufficient oxygen to their muscles where the fraction of oxygen in the air is only one-third that at sea level and where humans can take only a few steps before being exhausted?

The birds' lungs, with their impressive capacity to extract oxygen from the atmosphere, are only part of the solution. Next, the blood must extract the oxygen from the air going through the lungs and deliver it to the muscles. One of the major adaptations, as seen from differences between the goose and a domestic duck, for example, is that the goose's hemoglobin has a very high capacity to bind oxygen in the red blood cells. Thus the blood delivers more oxygen to the tissues per unit of blood pumped by the heart.

Next, the oxygen must pass from the blood to the muscles where it is used. In all animals with the capacity to take up and use oxygen at high rates, the muscle tissues are dark in color because of high concentrations of myoglobin, the dark red protein that takes up (binds) and thus helps remove the oxygen from the blood and deliver it into the cells, where the mitochondria, the tiny energy-producing batteries, utilize it by use of appropriate enzymes. Migrating geese have much higher concentrations of these enzymes for fuel utilization in their mitochondria than the nonathletic, sedentary domestic duck. All these various features working together contribute to the animal's $\dot{V}_{O2}$ max, or aerobic capacity, which limits sustained power output. In humans, $\dot{V}_{O2}$ max is an accurate predictor of middle-to-long-distance running performance, other things being equal.

But other things are seldom equal. Energy supplies must be used sparingly and carefully hoarded. In human runners, drafting, or running in the wind shadow of another, is a well-known strategy of saving energy. Birds do it routinely, especially the larger sociable species, such as geese, swans, and cranes, by flying one behind another to form large lines or *V*s. In addition to drafting, they avoid head winds and wait for tailwinds before departing. As with runners, pacing—to achieve maximum distance with least overall effort—is also very important for achieving maximum range. Vance Tucker's studies of flying budgerigars in a wind tunnel, in the late 1960s, proved the point.

Tucker successfully measured the rates of work output (from oxygen consumption) of bird flight at varying speeds by training his subjects to fly in place with a mask over their heads (to retrieve gas for measuring oxygen consumption) and against head winds of varying speed in a wind tunnel. Flight speed then equaled wind speed. Tucker's apparatus was essentially what a treadmill ergometer (to measure rate of energy expenditure) is to a runner, and his results revealed, surprisingly, that birds' power output in flight is not a simple function of flight speed. Flying at 20 kilometers per hour, his budgerigars were expending energy at near their maximum aerobic (that is, work) capacity, or a $\dot{V}_{O2}$ max of about 35 milliliters of oxygen per gram of body weight per hour. (Human $\dot{V}_{O2}$ is usually depicted in milliliters of oxygen per *kilogram* of body weight per *minute,* so by those units, for comparison, the birds' $\dot{V}_{O2}$ max is 583, not 35.) Their minimum energy expenditure during flight, of approximately 22 milliliters of oxygen per gram of body weight per hour, occurred at a much higher flight speed, of 32 kilometers per hour. To achieve flight speeds greater than 32 kilometers per hour, the birds' metabolic costs of flight again increased, ultimately reaching their maximum of 35 milliliters of oxygen per gram of body weight per hour at a flight speed of 48 kilometers per hour. Simple calculations revealed that, given this relationship, the maximum flight *range,* per given amount of flight fuel, could be achieved at the flight speed of about 40 kilometers per hour, which corresponds to an energy expenditure slightly above the minimum. Thus, speeding up in flight would get them there faster, but with limited energy supplies, they might not get there at all.

There are now extensive data on flight speed of migrants derived from radar observation. These data have been compared with calculated flight speeds and energy costs on the basis of wingspans and body weights. These studies show that most species fly to achieve maximum range, rather than flying at minimum energy expenditure. Exceptions often prove the rule. For example, during the breeding season, swifts often sleep in flight—they spend the night flying back and forth at high altitude rather than perching. During such resting bouts, forward flight speed slows down and energy expenditure approaches minimum, because covering distance no longer matters as it does during migration. Apparently bird migrants exert themselves to achieve maximum distance, not with the least effort at any one time, nor at the greatest speed, but with a *specific* effort that yields the longest flight range at the least *overall* effort for their distance. Distance runners must also search for, and find, that specific level of effort that is precisely suited to them and the distance traveled. Timing is important as well.

Small birds migrate at night, it is thought, for two reasons. First, it gives them time to replenish their energy supplies in the daytime. Second, it solves the problem of dehydration. The large amounts of metabolic heat generated by the work of flying can be dissipated, as in hawk moths and most other insects, by passive convection (unassisted heat loss to air) alone, without resorting to additional cooling by evaporating water. By flying at night, the birds can continue in their strenuous effort without stopping to drink. Given what the birds do, it is clear that for a record human ultramarathon performance it would be advisable to run on a cool night in order to conserve water.

Many birds cross seas and hot deserts, where drinking and refueling are not possible. They must carry all their required water and energy supplies with them. Fat, when burned as fuel, does release water as a by-product, so fuel and water supplies are interrelated. If rate of water loss is not in excess of metabolic water production, as it is usually in insects and birds, but not humans, then burning fat for fuel can also serve the need for water.

As already mentioned, long-distance bird migrants fatten up like butterballs just before embarking on their long journeys. Do human ultramarathon racers have to do the same? It depends on the distance and the rules. If we were to race intercontinental distances and the rules were that we could not at any time eat, then yes, the winners would be those who fattened up. The thin runners, who could run fast at the beginning, would not stand a chance of finishing. However, we don't race over such distances, and the organizers of our ultramarathon competitions ensure that there are refueling stations along the way. We can eat what, when, and almost as much as we want. That being the case, human ultramarathoners are better off lean. As long as we can refuel and rehydrate on the way, any extra weight we carry is a burden that slows us down. Most elite runners, men and women, carry at least several percentage points of body weight fat (1 to 6 percent), and if we could use it all, it would be sufficient to carry us hundreds of miles.

The common supposition that women could be better

ultramarathon runners than men because, on average, they have more body fat is false. On *average,* women are slower runners than men at all distances, and this sexual difference shows up especially at the longest distances. As I'll indicate later, there are biological reasons that relate this to genetic trade-offs. Some animal traits are sex specific. When women do run as fast and far as men (as many can), they likely do so at a reproductive cost. They must lose so much body fat that ovulation ceases. Animals are consummately pro-choice. Their bodies commit to the massive task of reproduction only when the resources to pull it off are available.

I'm unable to run hundreds of miles without stopping to refuel, despite my body fat, and I'm humbled by what is routine to any songbird or sandpiper. I'm awed by their ability to fly unbelievably long distances to and from specific pinpoints on the globe. Some might argue that if I were a bird, I would not be able to enjoy my fantastic annual journeys, following the sun from perpetual daylight on the high Arctic tundra to the pampas in Argentina and back again, but I think they are wrong. Birds are not likely driven by the logic of what they have to do. Instead, they are motivated by powerful urges. They behave in ways that feel right and pleasurable to them. Feelings of pleasure are a product of evolution that makes healthy organisms do what helps them survive and produce offspring, in the same way that fear makes them shy away from danger. What makes the blackpoll warbler strike out south in the fall after a cold front is probably not fundamentally different from what motivates me to jog down a country road on a warm, sunny day. We both respond to ancient urges that have adaptive roots. We have much in common, but our differences make direct comparisons of our endurance physiology difficult. With antelopes we're much closer.

NINE

The Antelope's Running Prowess

To the farthest limit he searches out. . . .
—JOB 28:3

Our world is full of countless mythical antelopes that one could race forever, chase, and never catch. Yet even those fleet-footed ones of hide, bone, flesh, blood, and guts that appear to be invincible are mortal. One legendary animal that seems ambiguous, hovering as it does in between the real that you can touch and the humanly unattainable, is the pronghorn antelope of the American plains. It has been clocked running at 61 miles per hour—almost twice as fast as a racehorse—and not just in a short sprint. It can reputedly cover 7 miles in 10 minutes. The Hopi tribe believed the antelope to be a spirit messenger and a powerful medicine. In a recent issue of the international journal *Nature,* the pronghorn was declared the world's premier ultrarunning animal, the best distance runner that muscle and bone and blood could produce. It is, of course, a mammal, and to the tiniest detail of our anatomy and biochemistry, we are also mammals. What exactly does the pronghorn antelope have that we don't, and how did it become such a superb runner?

For over at least 4 million years, the pronghorn (*Antilocapra americana*) has coexisted with predators on the open plains of North America. It has evolved in a habitat that offers long views in many directions, and it has survived not by hiding from its predators but by outrunning them. In turn, the predators—most likely saber-toothed tigers, wolves, cheetahs, giant short-faced bears, dire wolves, dholes, and hyenas—have had to come close to matching its running performance. In the "arms race" for speed and endurance that was waged for millions of years, the slowest pronghorn in many a chase got eaten. Weaknesses were exposed and culled out of the gene pool. John A. Byers, at the University of Idaho, points out that at the present time, "pronghorn are ridiculously too fast for any modern predator," and their running prowess is a "ghost" of that previous selective pressure that was greatly relaxed about ten thousand years ago after the late Pleistocene extinctions that decimated the North American fauna, at about the same time that humans arrived in America from Asia across the Bering land bridge.

One key to the pronghorn antelope's unique performance is its extraordinarily high maximum rate of power output (the previously mentioned aerobic capacity, or $\dot{V}_{O2}$ max). Rate of oxygen uptake sets a limit on the sustained exercise level. Experimentally, the maximum $\dot{V}_{O2}$, or maximum aerobic work capacity, is determined by gradually increasing the animal's workload (for example, by forcing it to run on a

Pronghorn antelope

treadmill at increasingly greater speeds or steeper inclines) until the rate of oxygen uptake reaches a ceiling. The workload can then be increased still further, but only for a very short time—seconds—at the cost of the buildup of lactic acid in the blood. In sprinters running all out, the quick buildup of lactic acid is subjectively felt as "dying," when the muscles seem to congeal. The extra oxygen that must be taken in after the exercise to oxidize the lactic acid produced during the exercise is called the oxygen debt. The increased speed bought by running for a few seconds beyond the $\dot{V}_{O2}$ max always costs dearly later. The cost of that speed is worth it only if the chase is reliably short and yields results.

In one study by Stan Lindstedt and colleagues, pronghorns were galloped on a treadmill at 10 meters per second wearing polyethylene masks to collect air for measuring aerobic work output. The incline of the treadmill was then increased to 11 percent, and the antelopes then registered an impressive $\dot{V}_{O2}$ max of 300 milliliters of oxygen per kilogram of body weight per minute. This is only half that of a budgerigar and a fourth that of a sphinx moth in flight, but it's impressive when compared to that of an elite human distance runner such as Olympian Frank Shorter: about 71 milliliters per kilogram per minute. Aside from sampling the air from the mask, blood was withdrawn from an artery within three to four minutes of running at $\dot{V}_{O2}$ max to examine blood gas content and also to check for lactate to verify that the animals were indeed running at their $\dot{V}_{O2}$ max and not beyond it.

Comparisons of an animal's relative power output must be made by taking body mass into account. Smaller animals generally have far greater rates of power output per unit of body mass, as the above comparison of a moth, a bird, and a human indicated. But taking the effect of body mass into account, the maximum aerobic capacity of the antelopes was still about three times greater than predicted for their approximately 71-pound body mass. Deviation from the predicted is called adaptive deviation, and the huge deviation (300 percent) in the antelopes' predicted $\dot{V}_{O2}$ max indicates that these animals indeed have unique athletic prowess with respect to aerobic metabolic specialization. (A sprinter's athletic prowess is the opposite—to run anaerobically, that is, without oxygen uptake, and to be *able* to accumulate a huge oxygen debt rather than to avoid one.)

What does the antelope get from its extraordinarily high $\dot{V}_{O2}$ max? The answer is, the same as human athletes do. A high rate of power output converts to the potential for high *sustained* running speed. For a given speed, however, the antelopes' oxygen uptake is similar to that found in other species. Antelopes' running speed can therefore not be explained by lower cost of transport (that is, great efficiency), and thus their high $\dot{V}_{O2}$ max corresponds directly to aerobic running performance.

The issue boils down to the question, what unique features account for the antelopes' high $\dot{V}_{O2}$ max, which is necessary to support their sustained high running speeds? To answer that question, researchers made comparisons with a similar-sized ruminant mammal, the goat. Goats are neither swift nor long runners. They are instead good climbers, and they evade predators not by outrunning them but by inhabiting inaccessible mountain ledges. Being safe up on ledges, they don't need to run, and they have only one-fifth the aerobic power capacity of the antelope.

In all structural aspects relating to physiology that were examined and that relate to rate of oxygen use, the antelopes were superior to the goats. Antelopes have more massive windpipes, three times larger lung volumes, greater gas diffusion capacity through lung tissue, an oversized heart, more cardiac output, greater amounts of hemoglobin concentration in the blood, more muscle mass, and greater numbers of mitochondria and hence more oxidative enzymes in their muscles. They also regulate their muscle temperature 2.6°C higher than goats do. If we recall the beetles' running speed as a function of body temperature, such a temperature difference should alone result in a 35 percent greater metabolic rate. In short, the antelopes' superb running capacity relative to goats' does not depend on any novel mechanism. There is no magic. Instead, the antelopes' unique capacity is achieved by enhancing a specific suite of many of their normal mammalian features. There are no tricks. No one adaptation by itself makes the difference. Pronghorns are just better at everything that affects sustained running speed. That is as expected. There is no use in having any one link in the system with extra capacity relative to any other—there is no point, for example, to have an oxygen delivery system to the muscles that greatly exceeds the capacity of those muscles to utilize the oxygen. Aside from that, the antelope is more than the sum of its parts, and no dissection will ever isolate the greatest of all antelope attributes, the antelope's spirit and enthusiasm for running. As Gary Turback says in his loving treatment of this magnificent creature:

> History is full of accounts of antelope apparently choosing to race a steed, car, train, or whatever just for the sport of it. . . . Is it too much to think when the prairie air hangs cool and crisp that pronghorns do not rise from their beds, stretch, and race off across the dewy morning grass at top speed . . . just because it feels good? Why would a sleek pronghorn in its prime not want the wind to whistle by its ears as it flies along at fifty or sixty miles per hour? Or want to hear the rhythmic pounding of its hooves on the prairie sod? Or to burst with bestial pride at being the best there is? Biologists may scoff at this, but as surely as sunrise, this happens. It has to.

Most biologists would not scoff at the message in this quote at all. Play serves a vital function in many animals. It serves the ultimate function of practice, and it is motivated by pleasure. Pleasure is a proximate mechanism for achieving many ultimate benefits.

John A. Byers, who has made a detailed study of play, has found that to the fawns of pronghorn antelopes and other ungulates that require speed to survive, play is fast running that may be interspersed with twists and leaps. It has long been argued that such exorbitant, apparently useless expen-

Young pronghorn at play

diture of energy is a survival cost. Contrary to this supposition, Byers found that those pronghorn fawns who played more had a greater chance of surviving the first month of life than those who played less. Similarly, carnivores' main play is mock predation, as it is in us, and it ultimately makes them better predators. We're the world's only hunters who will voluntarily handicap ourselves in the hunt if we are not dependent on the meat. We may hunt by choice using less effective weapons, such as antique guns and even bows and arrows, when we could use high-powered rifles with telescope sights instead. For us, hunting is not (now) always a necessity. It is also play.

If the antelopes' spectacularly high work capacity is adaptive (to them), why don't we find it also in the goat or in ourselves? Why shouldn't a goat be able to climb and run like an antelope? The answer evolutionary biologists usually give to this question is that everything has a cost. One potential cost of a high $\dot{V}_{O2}$ max could be that it results in an elevated basal metabolic rate. That is, having a large muscle mass packed with mitochondria might mean that (as in a car with eight cylinders that can never be shut off and must always at least idle, when compared to a car with only four cylinders) more is squandered in the long run. However, goats are not known to be picky or finicky eaters, and it turns out that antelopes at rest eat *less* food than similar-sized goats.

Another plausible hypothesis that has been proposed for the cost of the antelopes' high aerobic capacity, which is achieved in part by having a high percentage of metabolically active muscles but *little* mass of metabolically inert fat, is that this ratio makes them very vulnerable to food shortages. Like human marathoners, antelopes need to be and are very lean, and the antelopes' leanness or lack of energy reserves can sometimes have a high price. It makes them vulnerable to periods of cold and snow, when energy expenditure must rise for heat production but food may be scarce.

In the harsh winter of 1984, thousands of antelopes died in Wyoming. In the South Dakota winter of 1985–86, 80 percent of the state's fifty thousand antelopes perished from cold, heavy snow, and wind. And fences. The animals were unable to escape the weather by migration, and piled up dead in droves behind fences. They have powerful jumping capability, but they have not evolved, like forest-dwelling deer, the behavior of making vertical jumps. Although the task would be trivial for its body, its mind does not reach that high. The mind leads, the body follows. A pronghorn antelope can't conceive of jumping over a fence.

No studies are available, but I suspect that the antelopes

may be more compromised than is assumed with respect to ultrarunning endurance as well, and for the same basic reason that they are susceptible to periods of cold and snow. Pronghorn antelopes reduce body weight not only by having little fat but also by having a very small stomach—about half the size of comparable slower grazers. They are therefore compromised for endurance, because to continue to run fast for long durations they are forced to refuel at frequent intervals, and on high-energy-containing food at that. Pronghorns are picky eaters, choosing broad-leafed plants generally growing best where herds of bison have grazed off the grass.

Although the pronghorn antelopes, because of their exceptionally high aerobic capacity, have been touted as the ultimate mammalian endurance athletes, nobody has yet put the endurance of the pronghorn antelope to a serious scientific test. At least two human runners have tried. Dave Carrier, who works on the biomechanics of locomotion at the University of Utah biology department, and his brother, Scott Carrier, had heard tales of how in the old days Navajos and Paiute hunters had chased antelopes to exhaustion. The brothers, working as a team, tried to do the same, but Dave told me, "We failed miserably." The antelopes, who travel in groups, would dash off over a hill and "use the terrain to ditch us." The brothers would follow, and when they came over a rise, they'd see where the chased antelopes had joined others. By joining others who were still fresh, these animals were essentially using relays to outrun their pursuers. That may indeed be an antelope strategy that has worked in the past with pursuing wolves. (Humans arrived too recently in America to have been a strong selective pressure on their performance.) I suspect that a fair test of human versus antelope *running* performance would be to release a pronghorn in unfamiliar and open terrain, after painting it fluorescent orange. This might be a new and even more interesting running challenge for humans than a currently popular one of running around a quarter-mile track for 24 hours.

I consulted my friend the folklorist Barre Toelken at Utah State University (who had lived with and married into the Navajo tribe during the 1950s), trying to find out if the practice of running down deers and antelopes, in the days before long-range killing by rifles, might be referred to in folklore.

"I saw it done in the 1950s," he wrote me.

> But it's deer rather than pronghorns. What I saw was my friend Yellowman (in the 1950s he was about forty or forty-five) jogging along on the trail of a deer in semi-open desert country. The deer runs in bursts and then stops, listens, and then sprints again. The hunter, by consistently jogging along the trail left by the animal, eventually tires it out. Then, approaching the exhausted deer, he slowly puts an arm-lock on it and holds his hand over the mouth and nose of the deer, smothering it. His hand is supposed to have corn-pollen in it, which is considered sacred. The deer dies while breathing the sacred substance, and then its hide can be used as a sacred deer hide, unblemished because it's from an animal who was not punctured when it was killed. I never heard of the Paiutes doing it, but I know very little about them. I don't personally know of anyone who still does it, but there must be some people, since sacred hides are still in demand for ceremonies, and they're still obtainable. It usually took Yellowman all afternoon to run down a deer.

Yellowman, unlike most other predators, was undoubtedly not fooled by the deer's tactic of trying to avoid a race. Most predators are very selective and try to chase only what they stand a chance of getting in a *short* race. Deer exploit that tactic either by trying to get a headstart or by blasting off waving their white tail flag, thus showing they've got a headstart and a fast one at that, so that they won't be followed.

As I later learned—from *Indian Running* by Peter Nabokov (1981)—many Native American tribes highly prized running ability for hunting and for war. Many tribes had traditions of chasing down animals directly or indirectly on foot. In the Great Basin, antelopes were hunted after being chased on foot into *V*-shaped corrals. Before the Omahas of the central plains acquired horses, they had buffalo runners, who scanned the skies for ravens that signaled the location of bison herds, and then they ran back to the camp to recruit hunters for the attack. Hopi prized black-tailed jackrabbits and ran them down by following their fresh tracks in the snow. Both Pueblo and Yuki hunters ran deer to exhaustion, as did Papagos and Pimas.

These days, man versus animal races are sometimes devised mainly to settle bets. One of these that objectively measures the speed and endurance of humans versus another running animal is the annual "Man vs. Horse" race held at the Welsh town of Llanwrtyd Wells. This race is serious. It is sponsored by William Hill, the largest bookmaking company in Great Britain, which puts up £21,000 ($31,500) payable to the first runner who beats a horse. So far (after two decades) they haven't had to pay up; no individual runner has beaten a horse that is paced by a human jockey (relay teams of four runners do so often).

Nevertheless, the contest is close. In the last race, Mark

Croasdale, a British runner who has won the Marine Corps Marathon in 2:23, came within 80 seconds of the winning horse. Since the best human marathoners have run marathons in near 2:08, it is probably only a matter of time before one of them will beat the best horse in a long-distance race, even one as short as this one, which is four miles less than the marathon distance.

In summary, although pronghorn antelopes and horses are superb runners for moderate distances of maybe 20 to 30 miles (or possibly more), there is no evidence that they are ultramarathon distance runners. Antelopes and horses will never have to run 30 miles at a stretch from wolves. At least on the open terrain of Yellowstone Park, wolves usually catch elks within about a mile or so. If not, they give up the chase. According to Doug Smith—a wolf researcher and director of the Yellowstone Wolf Recovery Project, who routinely watches wolves hunt while observing from small aircraft—wolves seldom pursue for more than 2 miles. I suspect, therefore, that elks or pronghorns may perform poorly relative to a trained human who is motivated by the personal sense of accomplishment, such as that achieved by winning a race.

Speed is meaningless unless the distance is specified. This concept is best illustrated by human runners, among whom different running specializations are apparent. I plotted world record running speeds of both men and women (separately) as a function of distance from 100 meters to 200 kilometers. As expected, maximum running speeds by men (about 37 kilometers per hour) decrease dramatically with distances raced—about threefold as the racing distance increases two-thousandfold. However, the threefold decrease in running speed is not uniform over the whole distance. Speed decreases uniformly at first, but then plummets precipitously after about 1 kilometer; then it again decreases uniformly, before again plummeting precipitously after another break point, at about 30 kilometers. I interpreted the two transition points as a reflection of different physiological specializations, the first from anaerobic to aerobic metabolism, and the second from carbohydrate to fat metabolism. When I plotted the same curve for women, it showed the identical break points, which suggests the same physiologies as related to running specializations. However, at all distances, world record running speeds for women are slower than men's, especially at the longer distances. A March 2000 paper published in *Nature* confirms (or rediscovers) some of the same points made fifteen years earlier using a different data set.

Only forty years ago, there were hardly any competitive women runners. Now there are as many female as male athletes in many sports. Running is not just biological destiny. Rather, it is a biological capacity that is now largely a cultural phenomenon. Women, it turns out, are as eager and competitive runners as men are. Nevertheless, average differences in male versus female performance are clearly visible in the outcome of almost any race today, and those differences are reflected (though they need not be) in the records, the exceptional performances. However, no matter what the records say or what the population average may be of the group that we belong to, neither says anything about what individuals may be or could become or could do. Population data are required for formulating theories that are often perceived as laws of biology. But such laws don't dictate. They describe. Ultimately, the *empirical* performances of individuals create trends that are used for formulating theories, not *vice versa*. Each of us is unique and free to nurture his own gifts or dreams.

The reason for the *statistical* male running advantage is not known, although I suspect it may have something to do with average hip structure, tendency for weight distribution, and possibly different normal foot lengths. In any case, we all recognize the male-female difference, else there would be no separate male and female divisions and prizes at races. Despite the difference, women distance runners like Joan Benoit Samuelson, Uta Pipping, and Grete Waitz have bettered even Emil Zatopek's men's Olympic marathon record of 2:23, set in Helsinki in 1952. Men have bettered it also, by almost 16 minutes. Ann Trason wins ultramarathons outright.

Given that there are different physiologies for sprint, middle, and distance running, what would happen if the fastest men or women ultra-distance runners were to force an antelope or a horse to run 100 kilometers? Could the trained and motivated human force the animal to exhaust its fuel supplies? Until the race is run and the results are in, I'll reserve judgment on the pronghorn's prowess as an endurance athlete relative to both the best men and women ultraendurance athletes.

The antelopes, who have the equivalent of a V-8 engine in a VW frame, prove that aerobic capacity is absolutely crucial, as it is in human middle-distance runners, but I'll not be running a middle-distance. I'll be running an ultramarathon. Antelopes, like any runners, make compromises in energy reserves and in digestion in order to achieve speed in long-distance running. Those compromises are chinks in their armor that, in order to beat them in an ultramarathon race, would have to be exploited.

TEN

The Camel's Keys to Ultraendurance

Animals give us solutions to problems that are the product of evolution. They are the results of experiments that have been performed without bias or prejudice for millions of years. It is instructive to examine the results of these experiments, because there are millions of experiments that have yielded many very different solutions, and in this diversity we find both possibilities and pitfalls to specific objectives. Different animals have evolved to optimize different agendas that are not necessarily our own, nor ones to which we may wish to aspire. Instead of running, some evolved such heavy defenses as shells and spines (turtles and porcupines) or they have chemical defenses (skunks). In us, and presumably in other large vertebrate animals, including many dinosaurs, running speed is or was at a premium for both predators and prey. Since a camel's agenda is to travel long distances, camels provide us with insights into endurance running that antelopes can't.

Camels are not usually considered ultrarunners. Compared to antelopes, they are phlegmatic, ungainly ungulates. They've been of interest more for their marvelous ability to survive in heat and under severe conditions of desiccation. They show us how to handle an oversupply of heat with an undersupply of water. Nevertheless, their problem is precisely what ultramarathoners must frequently deal with. What do camels have that humans and many other animals lack? A wonderful book provides some answers. It is by Hilde Gauthier-Pilters and Anne Innis Dagg, who have observed and studied dromedary (one-humped) camels in the Sahara Desert for many years.

During any race of sufficient distance, all ultrarunners face the same problem, of managing overheating combined with fluid and energy depletion—the problem a camel faces routinely just to survive. The difference between an ultrarunner and a camel is that many camels face the problem almost chronically, while the marathoner has to confront it only for the duration of the race. A runner who races a distance of 1 mile or 1,500 meters would not have the problem at all, although overheating could begin by 10 kilometers. I would be running 100 kilometers, where the two other parts of the problem (fluid and energy depletion) are of critical importance. Camels routinely go far beyond 100 kilometers, and their example could be instructive because the extreme conditions most clearly illuminate problems and their solutions.

In contrast to antelopes and horses, camels are not fast runners over short distances. Their top running speed is reported to be about 10 miles per hour, whereas Secretariat, admittedly running a world record time, won the 1.5-mile

Camel

Belmont (in 1973) in 2:24, which averages to a speed of 37.5 miles per hour. Nevertheless, camels can reputedly go 100 miles in 16 hours. They can travel the 300 kilometers between Cairo and Gaza in two days. In a one-day race between a horse and a camel over a 176-kilometer course, the horse won, but just barely, since the horse died the next day while the camel kept going. Humans have run over 600 kilometers in four days and lived to run many more. Indeed, Yiannis Kouros, a Greek now living in Australia and possibly the greatest human ultradistance specialist of all time, has run 1,000 miles (1,600 kilometers) in 10.4 days, averaging 153.4 kilometers a day.

These human-camel comparisons do not apply directly, because the human racers were supplied with food and drink along the way, in any amounts that they wished. The camels received no such maintenance. Still, they provide us with a lesson: slow and steady wins the race.

Camels generally walk rather than run. When they run, their foreleg and hind leg on one side move forward simultaneously, and alternately with those of the other side. In this gait, the body swings from side to side, being supported first by the two legs on one side and then those on the other. This gait, which is opposite to our leg and arm movements while running, is economical for the camel, because it reduces the use of antagonistic muscles that would check the sway and provide for greater stability and maneuverability. Stability and quick maneuverability are not important for a large camel on the wide open desert spaces, where there is little danger from predators.

Camels are not indestructible. They may die from overwork or undermaintenance, as the French learned in the Egyptian Sudan in 1883 when they created a camel corps that took part in all their military operations until 1921. In 1900, twenty thousand of their thirty-four thousand camels died, apparently from overwork and improper care. Later, the French employed desert nomads as mounted camel men, and these *méharistes* had better success with their camels in the desert. The *méharistes* always possessed two camels, working only one at a time and resting the other. When traveling over long distances, they worked half the time. Yet the *méharistes,* typically led by a French officer, pursued such rebellious nomads as the tough Tuaregs for amazing distances. In one famous raid in March 1932, Captain LeCocq and his *méharistes* covered 770 kilometers in eight days while pursuing the emir of Adrar, who had killed some of their men. In 1911, Captain Charlet and his camel corps tracked the Tuaregs, who had been illegally trading in slaves, for more than 7,000 kilometers. After such long, ultramarathon exploits, the military camels typically needed six to eight months to rest, and badly overworked camels require a year's rest to recover fully.

When travel is fast and long as in the above routine camel treks, the high energy expenditure that is required promotes high internal heat production. Additional heat input from the external environment additionally threatens to overheat the animal. Physiologically the camel can counteract overheating by sweating. For a while. Since water is usually a limiting resource in the desert, the animal's endurance is most conspicuously focused on water economy. I suspect that an overworked camel that died in a race or on a long trek likely succumbed to the effects of dehydration. Perhaps not surprisingly, camels have evolved elegant solutions for water economy that reduce dehydration and its debilitating effects. These were not known until the 1950s, when they were discovered by Knut and Bodil Schmidt-Nielsen and their colleagues in a brilliant series of experiments. It was then already clear that the camel's legendary endurance was not attributable to diet. Camels make do with what food they can get. They eat practically any vegetation, even tough thorns.

For nearly two thousand years, it was presumed that the camel's secret to endurance was water storage. Pliny the Elder (c. A.D. 23–79) declared that camels store water in their stomachs, and this was long accepted as fact. A scientific publication in 1950 agreed with Pliny, going into some detail describing water sacs (glands with digestive fluids?) in the sides of the rumen (the first of two stomachs in ruminants). When the Schmidt-Nielsens began their studies to determine how camels could go for weeks without drinking in a hot desert climate (where a person would be prostrate in a day), they discovered no more remarkable water storage capacities than those of cattle, who need to drink daily.

Another myth that the Schmidt-Nielsens exploded with a few simple calculations was that the fat in the camel's hump is the key to water balance. The hump provides shading, and its fat contributes to water balance, but only in an indirect way. The 10-to-15-kilogram hump does indeed dramatically shrink in a starved and dehydrated camel, but that's because fat is used up as an energy source.

Metabolism of foodstuffs results in the production of carbon dioxide and water as by-products. The amount of this metabolic water, produced during the combustion of a given weight of foodstuff, is higher when fat rather than protein or carbohydrate is burned. At the same time, the oxygen required for any metabolism necessitates breathing, and breathing results in exhalation of air that is saturated with water. Under most dry desert conditions, more water is lost through the expired air than can be gained if fat were oxidized solely to produce water. There is thus no truth to the idea that camels metabolize the fat in their hump *in order to* get water.

The camel's hump is instead like a fanny pack that ultrarunners sometimes use when refueling stations are few and far between. It is not like a load of drink, but more like a load of concentrated food, like the commercial power bars that are currently popular. The advantage of carrying the fat on the back, rather than evenly distributed all over the body, is that it leaves the belly and other shaded areas less insulated and thus more available for heat loss from the body core. Perhaps even more important, the fatty hump serves, like our head hair, as a heat shield from the sun in the middle of the day, so that less water needs to be lost by sweating.

Part of the camel's secret is just plain toughness and the ability to survive desiccation. We're near death if we lose water equal to about 12 percent or our body weight, but camels can survive body water loss of 40 percent of body weight. After being dehydrated, a camel can ingest 20 to 25 percent of its body weight in one drinking bout. As in humans, the ingested water reaches the blood plasma from the stomach relatively slowly, requiring about an hour to attain a 25 percent equilibrium. But unlike humans, camels tolerate blood dilution to an extent not tolerable in other mammals. Our blood cells swell and rupture in dilution, and we can become very ill and even die from water toxicity if we drink too much liquid, especially when it is dilute (without salt or sugar) and therefore absorbed more quickly. There is no way to prescribe specific amounts of water, or concentrations of fluid supplement, to drink while running. The amount or dilution of water that is too much varies, depending on running conditions and on the individual. More commonly, runners drink too little and suffer from heatstroke. Camels suffer no ill effects from overdrinking; the camel's blood cells can swell up to 240 percent without rupturing.

In humans and camels, the body's blood plasma holds about 16 percent of the body water. When a camel is dehydrated to about 25 percent of its body water, the blood volume drops by only 1 percent or less, whereas in a human it drops more than three times as much, thereby thickening the blood, since the red blood cells remain. Blood that is thickened becomes viscous, like cold molasses; this greatly compromises its flow and strains the heart, reducing the blood's capacity to circulate easily and to carry heat to dissipate at the skin surface. Death from heatstroke becomes possible. The camel's erythrocytes (red blood cells) are uniquely oval shaped and small, reducing blood viscosity and permitting circulation through the capillaries despite fluid loss.

Water balance in camels is enhanced by their ability to reduce urinary water loss. We can produce only urine that is more dilute than seawater. In contrast, the camels' powerful kidneys can make the urine twice as concentrated as seawater, allowing them to get rid of much waste using little water. Camels can rehydrate even when drinking brackish or salty water that would dehydrate us since we'd expend water to flush out the ingested salts. Camels further reduce urea output (and hence the need to lose water through urine) by use of microbial flora in the rumen. Their gut microbes recycle urea wastes from protein metabolism back into protein that is later reabsorbed as a nutrient rather than being flushed out as a waste product. By a combination of these mechanisms, camels can save more water for sweating, and this enables them to travel farther in the heat.

In common with a variety of other desert mammals, one of the camel's main adaptations for ultraendurance on limited food and water supplies is its ability to regulate the rate of its metabolism through body temperature. The higher the body temperature, the higher the metabolic rate and the rate of heat production. Reducing both metabolic rate and body temperature is crucial when heat input is excessive and the body must hold the line at some critically high temperature that requires expending valuable water through sweat.

Desert animals in direct sunshine must resolve a paradox. They must try to increase metabolic heat loss, which favors reducing fur thickness, and they must decrease solar heat input, which favors increasing fur thickness to reduce heat coming in. They solve that problem by regionalizing insulation, having very thin or no fur in shaded body parts and very thick fur in areas most exposed to the sun. As already mentioned, the camels' solar heat input is reduced by the insulating hump and thick fur on the back. The surface temperature of fur on a camel's back may reach 70–80°C (158–176°F) in sunshine without harm, but that is because such temperatures are confined to where the sun hits the outside fur layers. Such temperatures cannot be tolerated by the skin itself, and left unregulated, bare skin would reach similarly high temperatures in the direct rays of the sun. Camels don't allow their skin surface temperature to go much above 45°C (113°F). There is only one means of reducing skin temperature: sweating, which in hot desert conditions is the primary water drain for camels and men. By comparing water loss of naked (shorn) and unshorn unaltered camels, Knut Schmidt-Nielsen found that in summer a shorn camel lost 50 percent more water than a fully furred one.

A major source of heat during exercise is, of course internal, that produced by the body's metabolism. Although air temperatures in the daytime desert may regularly be 40–45°C (104–113°F), there is usually a precipitous drop at night to the low 30s. Camels deprived of water take advantage of the night's low temperatures by allowing core body temperature to decline to as low as 34°C (93°F). In the day, they let body temperature rise up to 40.4°C (104.7°F). The low body temperature at night has the effect of reducing resting metabolic rate, and hence internal heat production. A relatively low body temperature in the morning also means that when camels start off, they go slowly but can travel for some distance before encountering elevations of body temperature that might require sweating or stopping. Further, the higher the body temperature they can later tolerate, the longer they can delay expending water for thermoregulation. (Hydrated camels, on the other hand, regulate their body temperature within the much narrower limits of 36–39°C [97–102°F].) The camel's apparent "inability to regulate body temperature," like a camel's slow running, was long thought to be a deficiency. Instead, both are elegant adaptations for ultraendurance on long treks through the desert. Whether their solutions to endurance and water economy are the result of previous exposure to heat and thirst, of genetic preprogramming only, or of both is not known. Our own body temperature, like the camel's, dips 2–3°F below "normal" at night, when heart rate drops as well. It takes most of us a while to warm up and get up to speed.

In summary, camels are able to cover long distances in desert heat because they are masters of water economy. They do so by starting slowly with relatively low body temperature, and tolerating dehydration and a high body temperature. They reduce their body temperature wherever possible by shielding themselves from the sun, and they have a remarkable series of physiological adaptations that minimize the use of water for waste excretion. Their blood chemistry favors tolerating dehydration when it does occur. Given the camel's example, an ultramarathoner who remains vertical in the heat should have long head hair or wear a hat, and shield his body with *loose* clothing. Frequent small drinks are better than tanking up, because we lack the camel's water-balance adaptations, being instead evolved for more speed at the cost of greater water loss. Beyond that, an ultrarunner would be ill advised to follow the camel's example and drink salt water, ingest huge amounts of freshwater at one time, or eat thornbushes.

ELEVEN

Athletic Frogs

In Mark Twain's famous story about the Calaveras County jumping frog, the frog's owner, the "infamous Jim Smiley," was challenged by a stranger who asked, "Well, what's *he* good for?" Smiley replied, "easy and careless," that "he's good enough for *one* thing, I should judge—he can outjump any frog in Calaveras county." Of course, we all know the outcome: the frog didn't budge, because when Smiley went out into the swamp to find a challenger for the bets that had been placed, the stranger stuffed the notorious jumping frog full of lead quail shot.

Since Mark Twain's time, frogs have budged a lot. When unencumbered, they're quite impressive at performing a rapid series of long jumps in a short time, that is, sprinting, using both legs at the same time. The current champion is a bullfrog named Rosie the Ribiter, who competed in the now-annual Jumping Frog Jubilee at the Angel's Camp fairgrounds in Calaveras County. She (?) set an all-time world record for a three-leap total, which is how the record is judged, of 21 feet 5.74 inches. That's not quite up to par with Bob Beamon's famous Olympic record of 29 feet 2.5 inches in the long jump event, but for a frog, covering more than 21 feet is impressive, even if it does take her three jumps.

A frog's leg muscles are designed for quick, explosive releases of energy. Frogs, just like sprinting cheetahs or humans, burn carbohydrates without the immediate use of oxygen, that is, anaerobically. Anaerobic performance never goes unpunished—in seconds the critter gets laced full of lactic acid and its muscles tie up. Undoubtedly, if Rosie had continued with more than three successive mighty jumps, there is good scientific evidence that each successive jump would have become very much shorter.

People do some pretty odd things to satisfy their curiosity. A colleague of mine at the University of California used to chase lizards on a miniature racetrack until they could go no farther. Then he grabbed them and ground them up in a blender and measured how much lactic acid they had generated. By assaying lizards after varying sprint durations, and at varying postsprint intervals, he was able to determine that it takes some lizards an hour or more to get rid of their lactic acid load. Frogs operate under the same constraints, and air-breathing divers who stay underwater without access to oxygen suffer the same lactic acid accumulations. There is no grand discovery here—just a reaffirmation of what we all know from personal experience. We can't sprint or hold our breath if we want to go far. Sprinting in the middle of a race is not to be recommended. The sprint must come at the end, because you can pay off the oxygen debt when you're done.

Some frogs, unlike Rosie, are ultraendurance athletes par excellence, and they annually stage their own remarkable

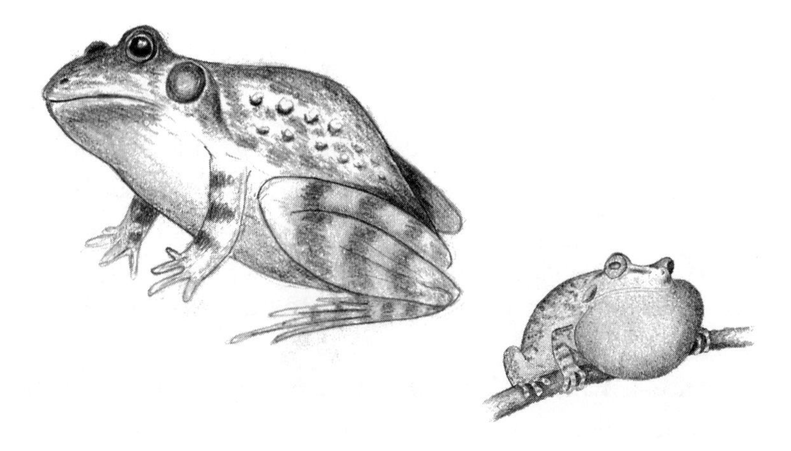

Left: Bullfrog   Right: Male tree frog calling

contests. These contests are performed strictly by males, and the contested skill is in aerobic shouting, not anaerobic jumping. The prize is copulation with female spectators. The longer a frog can sustain energetically demanding aerobic shouting, the better his chances of getting a prize. Naturally, each male puts out his best effort. The contests are not as open to the general public as the bullfrog jumping contest at the Angel's Camp fairgrounds in California, because they involve small, highly camouflaged individuals that gather only in or near isolated swamps and that usually don't reach their full stride until well after dark. The contests also have been staged at the University of Connecticut, where in the labs of Theodore L. Taigen and Kentwood D. Wells, details were divulged that would otherwise have remained secret. One of the first findings these researchers made was that the chorusing of male tree frogs (*Hyla versicolor*) is achieved at about 60 percent of their $\dot{V}_{O2}$ max, which is close to that of ultrarunners on long races. The frog chorusers maintain their extremely high work outputs for hours each night, apparently until exhaustion sets in near morning. Not surprising, not all contestants last the whole night.

In frogs of some species, there are thousands of competitors per contest, but individual male frogs usually pursue their strenuous efforts out of visual contact with their competitors. However, Taigen and Wells found that as soon as a frog sees a female spectator, he increases his vocal output to near 100 percent aerobic capacity. He can't keep it up for long, though.

Since females are attracted to and approach the most energetic callers, for a strictly short-term strategy a male frog who sees a female should call as vigorously as he can. But in the dark, he never knows whether or not a female is near. Therefore, when calling in the dark, he must be hopeful and call for as long and vigorously as possible. Those frogs with the highest aerobic capacities, who can keep up the most vigorous calling for the longest time, are the ones to leave the most descendants. Frog behavior and the anatomy and physiology that is the basis of the behavior are the results of millions of years of evolution, and the optimum results in the conflicting trade-off between maximal energy output and endurance are available for inspection.

The structure and physiology of male frogs is different from that of females, who do not enter the energy-demanding chorusing contests. Both males and females are of equal length and have the same leg muscle mass, but males on average weigh 1.25 grams while females weigh only 1.05 grams. This difference in mass is mainly due to highly hypertrophied body-trunk muscles in the males (0.18 grams for males versus 0.03 grams for females). The males' trunk muscles are so well developed because they are used to drive air over the vocal cords to produce what are very loud sounds for such tiny frogs. Without their muscle hypertrophy, the males presumably would be able only to whisper rather than shout. The females are silent. Similar sexual dimorphism in behavior and associated muscle anatomy and physiology exist in katydids and crickets, which pursue a similar mating strategy.

The frogs' trunk muscles, in contrast to the leg muscles, are uniquely adapted for aerobic metabolism. These muscles, like those of geese, antelopes, and human distance runners, are packed with mitochondria, the small power packs in cells within which all aerobic metabolism occurs. In the frogs, the mitochondria contain citrate synthase, a key enzyme for aerobic metabolism, in higher levels than is found in any cold-blooded vertebrate so far examined. The mitochondria in the trunk muscles of male as opposed to female frogs also contain almost twelve times the activity level of the key enzymes phosphofrustokinase and ß-hydroxyacetyl-CoA dehydrogenase for fatty acid metabolism. The conclusion can be drawn that, just as for human endurance running, aerobic fatty acid oxidation plays a key role in the endurance energetics of the frogs' chorusing. As in all other animals, physiology is closely tied in with behavior, and the frogs' behavior has evolved to maintain the highest possible rate of activity for very long durations. And as in ultramarathon running, that involves pacing.

In the laboratory, calling rates offer a direct measure of aerobic energy expenditure. In the field, one can estimate energy expenditure simply by measuring calling rates, much as one can deduce human energy expenditure on the track from running speed, after one has measured it on the laboratory treadmill. Comparisons of estimated aerobic output in the field with maximum rates observed in the laboratory make it possible to calculate the percentage of maximum effort that the frogs put out at any one time. This averages, as mentioned already, close to 60 percent, but the frogs start out their shouting matches at a much more leisurely pace.

Male tree frogs, like many ultramarathoners, who also pace themselves by starting out slowly, begin their evening chorusing contest at about six hundred calls per hour, then increase their calling pace (depending on the individual) gradually during the next two hours. They then gradually slow down during the final, predawn hour or so, when many individuals begin to drop out of the chorus. Taigen and Wells ground up whole frogs to measure body lactic acid accumulation and showed that even though the frogs' initial calling pace is slow, the lactic acid concentrations found in the first half hour are higher than those found later, at peak calling frequency! These results strongly suggest that these animals need a long period of warm-up, when glycogen is used as a fuel, before a switch-over to fat metabolism occurs. This is also the case for locusts during flight, as well as for human distance runners. The lesson to be learned from frogs is, start slow and work into a pace slower than the final pace.

The frogs' pacing involves not only starting slowly, but also varying the lengths of repetitive work bouts (each call) and rest periods (intercall intervals). Many ultramarathoners who race for 24 hours or longer ask themselves whether it is better to walk for 1 mile out of every 10 while maintaining the same overall pace, or to walk a tenth of every mile. The data from the frogs' calling behavior may provide a clue.

Male-male competition brings out profound changes in calling behavior, as does the presence of females. The frogs give either long calls or short calls. Females prefer males with the longest calls. Males chorusing in crowds, or those who are duped with playback from a tape recorder to simulate nearby calling males, give calls about twice as long as isolated males, who give primarily short calls.

Long calls are energetically more costly to make than short ones, but as the individual call lengths increase, the frogs adjust by reducing the call rate so as to keep the energy expenditure about the same. These results are not what one would predict from evolutionary logic, because *if* it costs no more to give the more attractive long calls than the less preferred short calls, then why make any short calls at all? Is there a penalty for making the long calls? There is, indeed. The penalty is, quite unexpectedly, a reduction in stamina. Even while calling at *given* rates of energy expenditure, males making long calls had dramatically lower endurance. For example, those frogs giving calls with durations averaging 350 milliseconds called for 3.75 hours per night on average, while those whose call durations averaged 500 milliseconds called for only 2.25 hours.

To date, nobody knows why making longer rather than shorter calls, at a given rate of energy expenditure, results in a dramatic reduction in the frogs' stamina. Can it be related to glycogen depletion? From studies in human runners, we know that although the fuel for long runs is primarily fat, we still "hit the wall" when muscle glycogen (a carbohydrate) has been depleted, even though fats may still be plentifully available. One hypothesis is that carbohydrates and possibly proteins help replenish the constituents of the central biochemical cycle, the so-called Krebs cycle, so that it can continue to burn fat. I suspect that making the long rather than the short calls results in a more rapid depletion of muscle glycogen because each *individual* call, though less than one second in duration, is a very intense energetic effort that favors the utilization of carbohydrate over fat. Possibly, slightly more muscle glycogen is used during the production of long as opposed to short calls. Glycogen is wholly or partly restored in the short intercall interval. By prolonging the call duration, there might be a shift in the equilibrium that results in faster depletion of critical glycogen stores, with the cumulative effect over many thousands of calls being that glycogen is drained sooner. As a result, the duration of exercise is reduced, even though the fat reserves are still abundant.

I wondered if the same principle might apply to human runners taking short rather than long steps. Most ultrarunners take short steps, and I would do so as well; long steps cover ground faster but tire one out more quickly.

It is still too soon to come to conclusions, but if I were planning strategy to set a record in a six-day race during which it would be necessary to alternate walking with running, I would shorten my stride and make my run/walk intervals very short rather than taking long runs and long rests. Obviously, I have no idea what constitutes "short" or "long" intervals. That could only be determined empirically, because we do not have the data on humans, much less specific individuals, to come to conclusions. Nevertheless, ultramarathoner Kevin Setnes told me of adapting a precise walking/running routine into his 1993 Olander Park 24-hour-run championship race, in which he extended his personal record by 35 miles and set an American road record of 160.4 miles. When I asked about his strategy, he told me his schedule of alternating walking with running was "the single most important factor in achieving that total [mileage]." He admits that he got his racing idea from a preliminary article I had written about frogs for *Ultrarunning* magazine, after I had read Taigen and Wells's original work on the tree frog. Maybe I should take the same advice. But I can't.

In Chicago, given the time in which I want to finish my anticipated 100-kilometer run, I will not be able to stop for even one second. The 100-kilometer race is like the 100 meter is to sprints and the 10 kilometer is to middle-distance running. It's the premier international ultrarunning standard. Nevertheless, it is still far too short a distance to permit any stopping if you are planning to win or set a record. It is a fast-paced race every step of the way. My pacing in that race will have to refer to overall running speed and possibly stride length, not to schedules of stopping and starting. Such running may not seem natural for humans now, given our very recent lifestyles. What most of us do now may not be a good indicator of what we had to do millions and at least many hundreds of thousands of years ago and still can do, given the right conditions. We can't be sure what that was, but as in other animals, our bodies still yield clues of what shaped us. They show what is still possible for us now, given the appropriate conditions.

TWELVE

To Run on Two (or More) Legs

Human ingenuity may make various inventions, but it will never devise any invention more beautiful, nor more simple, nor more to the purpose than Nature does, because in her inventions nothing is wanting and nothing is superfluous.

—LEONARDO DA VINCI, *fifteenth century*

A human runner has to make many strategic decisions. Some of those choices can be based on science, and he can look to animal models for guidance. In the same way that a painter must know the technical effects of color combinations, techniques of paint application, shading, and highlights, a runner must acknowledge physiology, the medium through which excellence is exerted. Nevertheless, although animals can reveal mechanisms, our performance, whether it is in a painting or in a race, is ultimately art because there is so much that varies.

Roadrunner

Of course, one thing we have no control over is how many legs we have. Yet leg number varies widely between different animals, as does running speed. Since running has evolved independently in various evolutionary lines of arthropods, dinosaurs, birds, and reptiles, and in various groups of mammals, we might reasonably ask if leg number affects speed. In the fourth century B.C., Aristotle said, "If one way be better than another, that you may be sure is nature's way." But nature's way is diversity. Aristotle made many original observations of nature, but he apparently did not know about historical constraints, or the compromises, which we must accommodate or adjust to.

We evolved from lobe-finned ancestors that crawled onto land with four limbs, with which we were blessed, or saddled, in subsequent evolution. Does multipedalism increase, or decrease, speed? In the arthropods, there is much variation in leg number and that allows comparisons. Millipedes, depending on species, have about a hundred or two hundred legs, and they are a study in slowness, even with most of their legs working all-out in waves, some going up and some going down at the same time. Centipedes, with only about fifty legs, are also not speedsters, except with respect to millipedes. Many of a centipede's legs are expendable and illustrate a novel way to use legs for a getaway. One type of centipede and daddy longlegs (an arachnid) drop off legs when chased, and the loose, wiggling legs twitch and distract the attacker while the owner runs off. Spiders, with eight legs, are faster than centipedes, and they grow a new leg for every one they lose. Perhaps they need them all for web building, as well as locomotion. Insects have six legs, and some of them, like the tiger beetles I mentioned earlier, are superbly fast runners, at least on a hot day. Others are laboriously slow.

For efficiency and smoothness of stride, few insects can compare with some species of cockroaches. Considerable progress has been made in elucidating how these insects run. As revealed with high-speed cameras, the champion runner, the American cockroach, *Periplaneta americana,* raises three legs at a time and keeps three on the ground. The first and third on one side and the second on the other are used as a unit. The roach moves using such alternate tripods. The difference between walking and slow running is simply the rate at which successive tripod steps are taken, although when really cruising, some cockroaches do something different. They spread their wings, shift their body weight to the rear, and become bipedal by running on their hind legs. American cockroaches can sprint this way at some fifty body lengths per second. By that measure, they run about four

Basilisk lizard (crested water dragon)

times faster than a cheetah, the world's fastest land animal in terms of absolute speed.

There are great differences between species of cockroaches. David George Gordon, who has written the world's most authoritative guide to cockroaches, points out that the German cockroach "is having a great day if it can go faster than a foot per second." The Madagascan hissing cockroach is even slower. It is a lumbering beast that nobody has yet had the interest (or patience) to time. Arwin Provonsha, when referring to American cockroaches, says, "These puppies are born to run." He should know. Provonsha is the curator of insect collections at Purdue University and the announcer at their annual All-American Trot, which features mostly American cockroaches. The All-American Trot, where such matters are brought under objective scrutiny, features cockroaches' footraces on a custom-built circular track with racers coming from entomology department research stock. Such pedigreed individuals lovingly named Hot to Trot, Sewer

Sam, Plain Disgusting, and the like are marked on the back with bright acrylic colors for the spectators, more than seven thousand in 1995. What induces the contestants to run is seeing the light of day; they are kept in the dark until the starter's gun. Ultimately, the contestants are running for their lives, because throughout their more than 500-million-year evolutionary history, the cockroach that could not quickly scoot into a dark hiding place was a dead cockroach. On the other hand, if you're big, heavily armored, and have formidable defenses, like the Madagascan hissing cockroach, then you don't need to run fast, and probably won't. (In the All-American Trot, the Madagascan species' talents are harnessed not in straight-out footraces but in pulling miniature green-and-yellow John Deere tractors.) This cockroach racing stuff has its serious side, too. There is the betting, of course, and the discovery that the ultimate in cockroach running speed is achieved by bipedalism.

All of the great quadrupedal dinosaurs were probably slow or only short-distance runners, but the bipedal ones (gallimimus, compsognathus, velociraptor, and others), according to paleontological evidence, were speedsters. The ostrich, a bipedal descendant of dinosaurs, is a superbly graceful runner that cruises at 70 kilometers per hour and can keep it up for long distances. Similarly, although some present-day lizards can run well on four legs, some species—the basilisk, the crested water dragon, and others—achieve their full running speed only by rearing up onto their hind legs. By switching from the quadrupedal to the bipedal gait, the basilisk can even achieve high enough speeds to run on the water surface, hence its name "Jesus lizard." If cockroaches and lizards achieve faster running speed by becoming bipedal, then it seems plausible that our own evolution, from semiquadrupedal ape ancestor to bipedal human, also had implications with respect to running speed. Certainly not everything, but something.

Bipedalism in mammals is associated with relatively open arid environments where long-range vision and rapid movement are both at a premium for foraging and predator avoidance. The bipedal mammals that come quickly to mind include kangaroos in Australia, springhares and early hominids in Africa, kangaroo rats and jumping mice in North America, and gerbils in Asia.

All of the bipedal animals that run fast do so by a rapid succession of long leaps, either alternating between legs or kicking off with both legs at the same time. There is considerable heavy impact of the feet striking the ground, and with that impact comes a potential loss of energy. However, mechanisms have evolved to harness some of this otherwise wasted energy. It's in the anatomy. As the foot is depressed on landing, the heel (Achilles') tendon is stretched, and when the foot rebounds with liftoff on the toes, the just-stretched tendon, or springing ligament, contracts and releases stored energy. Up to 40 percent of the energy absorbed by the impact is retained in this ligament, to be returned to the body during the second step. The arch of our foot also depresses and stores energy, and experiments with human feet from cadavers suggest that up to 70 percent of the energy that goes into a depressed foot arch may be returned as well (although the elasticity of muscles and tendons that gives us bounce greatly decreases with age). Obviously, running surface also makes a huge difference, a fact well known to track runners. Runners reach their greatest speeds on tracks that compress 5–8 millimeters (about the same as the compliance of the spring in the foot arch), and experimental tracks of varying stiffness can return 90 percent of the energy stored in them.

Track shoes do the same, but the compliance of the shoe has to be tuned to the running surface so that the shock of the step is not merely absorbed and the energy dissipated. Using good bouncy shoes and simultaneously running on a springy track does not add up to getting more energy back from each step than we put into it. To the contrary, energy cancels out. It's like bouncing a ball. A rubber ball will rebound higher from a hard surface than from a surface that yields.

Given our foot design, barefoot running on the right surface can be an efficient way to race, if the soles of one's feet are tough enough so that one can strike one's feet solidly enough against the hard ground to generate rebound. As mentioned before, I'd already experimented in Africa, and I found out that my feet weren't nearly tough enough. Those of Abebe Bikila were, when he ran barefoot and won the 1960 Rome Olympic marathon race in record time of 2:15:16.2. However, he subsequently ran more than 4 minutes faster in the next Olympics, at Tokyo, this time wearing running shoes. As far as I know the subsequent sub-2:10 marathoners have all been shod.

From beetles, cockroaches, ostriches, and cheetahs, evolution of greater running speed has been associated with a reduction in foot weight, achieved by reducing the number of digits, and with a *lengthening* of the foot and toes. This trend is best seen in the evolution of the horse. Horses run on the tip of just one single greatly strengthened and elongated toe on each foot. Ostriches also run primarily on one enlarged toe, with a second, smaller one providing lateral support. Deers and antelopes run on the tips of two toes, but the metacarpals of those two toes have become fused to form one elongated bone. A greatly elongated foot makes the leg lightest toward its terminal end, because the major muscles that power the leg are located high up near the trunk, being attached to the foot with long tendons. That arrangement not only helps to lengthen the stride, it also makes each stride energetically less costly since a light leg can be swung forward and backward faster and more easily than one weighted at the end.

Our feet are uniquely adapted for running relative to our cousins the apes', whose feet have five digits still fully functional for grasping. Having given up easy climbing for fast running, we have toes that are now almost useless; we cannot use them for grasping, and in order to achieve top speed, we run "on our toes," with most of the power during the step's kickoff on the run being applied through the big toe. When we sprint, the hind part of our foot barely touches the ground, thus effectively lengthening the leg. Power comes from the front end. For all practical purposes, all of our toes could as well be fused, or our large toes could be enlarged and the others deleted, if we were uncompromisingly designed to be pure sprinters.

We don't know which variables evolution might alter to achieve greater speed versus endurance. Undoubtedly they would be many. But if the examples of other animals is a guide, and if we were subjected to a few million years of strong selection specifically for running speed, then evolution would undoubtedly alter our feet! As I have indicated elsewhere, most women runners—even the elite—have slower running speed than men. No definitive explanation exists to account for this difference. Could foot length be a factor? Women's feet are shorter than men's, and in one informal survey that I did, I discovered that they were possibly even shorter than predicted on body size alone. Did males face stronger selective pressure than females to enhance running speed?

To some animals, legs can be a handicap. One very fast vertebrate land animal that I encountered on the African acacia steppe has no legs at all. Being young and foolish, I wanted the skin of this exotic animal as a trophy. I'd chased my quarry under a bush in short, dry grass. As I got close to try to slay it with my shotgun, I saw a quick movement and a set of beady coal black eyes. In almost the same instant, the creature lunged out at me. I jumped back and began to sprint. Glancing back, I could see it right at my heels. Running fast, I soon reached a clear sandy patch, at the edge of which the spitting cobra stopped, reared itself up three or four feet, and again glared at me. I turned and shot. After my sojourn in Africa, my Maine teammates joked that the snake episode had taught me how to run. Perhaps it did in the ultimate, evolutionary sense. We all learned the virtue of speed that way.

Cobras are long, thin snakes, as smooth to the touch as polished glass, all the better to slither fast. Snakes move forward much like fish do, by applying lateral force against a medium while being slippery to it. It's basically the same principle as that used in sailing, where wind rather than muscle power provides the energy for forward momentum. Caterpillars appear to run, but they don't do it with their legs, either (which are restricted to three very short pairs at the front). Caterpillars move forward, like seals humping along on land, by a series of posterior-anterior peristaltic waves. To imagine how they move, think of a hot dog with semiliquid contents contained by an elastic, semirigid shell. Internal muscles contract in series from tail to head, and as a contraction wave passes any one point of the body, that portion is lifted off the ground and is telescoped forward for a "step" as the caterpillar extends. Speed is largely a function of "stride" frequency, not stride length. The caterpillars' legs hold and anchor, but at this point in evolution they no longer have anything to do with their original role in powering the locomotion. Their legs show us what they once did. The ancestors of snakes also had legs that ultimately became useless, if not an impediment to locomotion, yet some of them still have vestigial leg elements internally.

In humans, morphology also still gives us clues of our ancient history. As Charles Darwin stated in *On the Origin of Species,* "Organs now of trifling importance have probably in some cases been of high importance to an early progenitor, and, after having been slowly perfected at a former period, have been transmitted to existing species in nearly the same state, although now of very slight use." An animal's musculoskeletal system provides clues to selective pressures that have acted on the organism. Similarly, our nervous systems and our basic behavioral tendencies are just as much products of natural selection as are our muscles and bones.

Morphologically and behaviorally, we reflect our past. For us to run well requires not only an efficient bipedal running form, but also elastic Achilles' tendons, strong big toes, and perhaps even more than anything else, special psychological tendencies. I will propose a hypothesis of what those psychological tendencies might be and how they could have arisen in our prehistory as apes on the African savanna.

THIRTEEN

Evolution of Intelligent Running Ape People

When you experience the run, you . . . relive the hunt. Running is about thirty miles of chasing prey that can outrun you in a sprint, and tracking it down and bringing life back to your village. It's a beautiful thing.

—SHAWN FOUND,

the 2000 American national champion at 25 kilometers

The apelike creatures that were our ancestors were a strange breed. They were perhaps at first awkward scavengers on the African plains, who later became bipedal predators. They were neither big nor swift and had to make up for it with sociality and smarts.

The above scenario of our evolution as bipedal savanna hunting apes and ultimately people is like a large house with many rooms in various stages of construction, from rough to nearly finished. It is the result of constant restructuring and elaboration by many builders of a great range of expertise. The various parts were contributed or built by paleontologists, anthropologists, behavioral field biologists, ecologists, physiologists, and anatomists. I will here try to show some of the evidence and logic for the construction of the main frame of the house that contains so much. I will then explore what I think are ramifications relative to our psychological and physiological capacities as endurance predators. In this limited space I cannot here argue all the pros and cons of each specific point. I can only review the scenario as it seems to me to make the most sense. And a central part of that scenario is, I think, our endurance. Furthermore, the key to endurance, as all distance runners know, is not just a matter of sweat glands. It's vision. To endure is to have a clear goal and the ability to extrapolate to it with the mind—the ability to keep in mind what is not before the eye. Vision allows us to reach into the future, whether it's to kill an antelope or to achieve a record time in a race.

Our specialty as bipedal runners spans a history of at least 6 million years. It probably began in Africa, when open or semiopen plains were replacing forests and our ancestors began to diverge from other apelike creatures to venture out of the forests and feed on the vast assemblage of herbivores supported by the growing seas of grass. There were many other predators out there and little safety from them in trees. Nor was it easy to hide.

Life on the plains generates arms races between predators and prey. Here we find such sprint champions as the cheetah and the various species of antelopes it hunts. Also on the plains were (and still are) such cooperative predators as pack-running canids and hyenas, which catch prey by capitalizing on the weakness that sprint speed produces, namely compromised endurance. In turn, the sprinting prey species sought and found some measure of safety in numbers. Antelopes are consummate herd animals.

The very first bipedal hominids were undoubtedly not superb runners and they needed alternatives to raw running speed for survival. They would have cooperated to hunt, as some monkeys and apes do today. On the plains, even some normally solitary predators became social in order to hunt. Lions, who live in groups unlike all other felines, are a prime example.

Speed was useful, and necessary as well. We'd never run 60 miles an hour like a cheetah, but a cheetah doesn't need to run for an hour. It needs to run only for a half minute, and it can't run much farther before it runs head-on into the problem of overheating and lactic acid buildup, and must stop. The speed-disadvantaged hominids had other advantages beyond their already existing sociality. Not only did they have grasping hands useful for climbing and throwing, and ultimately for tool use, plus an edge on intelligence, they also developed running endurance while remaining upright.

Human bipedalism in running has been thought enigmatic because it is energetically expensive relative to quadrupedal running. Nevertheless, when used for traveling long distances on the plains, bipedalism was likely a great improvement over the knuckle-walking of hominids' ancestors. In evolution, almost every solution is the result of compromises. Energy efficiency was sacrificed in favor of freeing the hands for other use. For instance, hands were useful not only for throwing rocks and sticks, and later for making, carrying, and using weapons, but also for carrying our babies and prey to our safe campsites. Our progenitors, like chimps today, could likely hurl objects. By standing upright, they could see farther and defend themselves, when necessary, in several directions at once.

British physiologist Peter Wheeler has proposed that our bipedalism evolved in part for thermoregulation under exposure to the blazing tropical sun. As the examples of the hawk moths, bees, and camels demonstrated, reducing heat input or increasing heat loss translates to greater endurance. Wheeler photographed a model humanoid in either bipedal versus quadrupedal posture and found that in the bipedal posture it experienced 60 percent less direct solar radiation. In addition, in that posture the body is better situated to take advantage of breezes for convective cooling. As I've discussed (Chapter 12), bipedalism can enhance speed, but even if it did not, and even if it is more energetically costly, it was still a better bargain than giving up our tool-using hands, reducing our visual range, and compromising our endurance in the heat. So, on balance, human bipedalism is not enigmatic at all.

The hominid line to which we belong likely diverged from apelike creatures about 5 to 8 million years ago. The first fossil traces of that hominid line have been found in 4.4-million-year-old rocks from Ethiopia that contain a creature called *Ardipithecus ramidus*. The australopithecines, or "southern apes" (after their discovery in Southern Africa), which were derived from *Ardipithecus,* were small brained relative to us, but as determined from skeletal remains and from footprints, they already walked upright. *Australopithecus afarensis,* of Lucy fame—the three-and-a-half-foot-tall female, was discovered in Ethiopia in 1974—is one of the best-known australopithecine fossils found. Australopithecines were bipedal intermediaries between the apelike and the subsequent *Homo*-like forms and were unlikely to have been able to outrun most large predators. Australopithecines needed other defenses. The endurance running capability that subsequently evolved in humans and that was derived from australopithecines must have been under some selective pressure other than avoiding predation.

Most likely, the australopithecines diverged from forest dwellers to occupy the dangerous plains not to avoid predation but to seek food there, despite, and perhaps because of, the predation there. Meat was in abundance on the plains for those who could catch it, for those who could take it from such other carnivores as leopards, cheetahs, and lions, and for those who could compete for it against hyenas, jackals, and vultures.

Given the reasonable assumption that the australopithecines were group-living as most present-day apes, it is not difficult to envision a plausible scenario for how they got their food. Traveling in groups and coming upon a predator-killed carcass, they might have chased off the owner with sticks and rocks. Takeovers would have been difficult at night, and it would have been easiest in the middle of the day when the predator had retired into the shade, leaving carcasses untended or at least less vigorously defended.

Intelligent hominids would have quickly learned how to find carcasses. Some years ago, I dropped off a dead horse near my mother's house in Maine to feed ravens. Her two dogs found the carcass with the raven crowd. Ever since then, the dogs have become eager raven watchers. Early plains hominids would have been no less able than circling vultures, my mother's dogs, and myself to recognize and heed signs of a recent kill.

On the African veldt, most predators need to kill frequently, because what they don't eat almost immediately is consumed by scavengers or spoils quickly. There is great competition for scavengers to be there first after a kill has been made, and the fastest come on the wing. In an intact northern ecosystem, Yellowstone Park, it is the same, only there the scavengers that come within a minute or so of wolves' making a kill are ravens, not vultures. Eagles, bears, and coyotes then use the ravens' activity as a cue that indicates the kill, and they also rush in. In Yellowstone, within about seven hours after wolves kill an elk there is nothing left but the bones. In Africa even the bones are eaten (by hyenas), and the carcasses are consumed even quicker.

During my year in Tanganyika (now Tanzania), I found one morning an unattended freshly dead cow in a deep streambed. By noon, when I came by again, there were well over a hundred vultures there feeding, and more were still streaming in all the time from all directions at once. Vultures, too, find carcasses by watching others, and their mobility allows these birds to exploit the diurnal niche of predator-provided meat. Similar competition for predator-killed carcasses would have existed on the grass savannas where our anthropoid ancestors evolved. Then as now, traveling fast and long would have been a great premium for getting to predator-killed carcasses before the competitors devoured it. Ultimately, the hominids' mobility in the heat could have been transferred to getting their own fresh meat, by hunting.

Although the earliest australopithecine-like hominids were probably not swift enough to run down healthy adult antelopes, there were undoubtedly many advantages for them to become ever faster. The races against scavengers and against others of their own species, that is, their closest competitors, could have become the bridge to races with formidable live prey. Once hominids were fast enough, they potentially could run down such weaker prey as calves, the old, and the injured.

Ultimately, what early hominids did routinely may have been less significant than what they could do in the times of greatest need, such as when no dead or injured animals were available. Running ability would have become ever-more valuable on the plains after meat became an important part of the diet. By about 2 to 3 million years ago, the bipedal plains hominids already had a leg and foot structure almost identical to our own. The fossilized footprints that Mary Leakey discovered show that they walked like us. It is reasonable to suppose that they could also run even before they evolved to several species of *Homo,* of which *H. erectus* was the first to leave Africa.

Another theory, one recently proposed by Richard W. Wrangham and colleagues, is that the big evolutionary change from australopithecines to *Homo* occurred after the invention of cooking, primarily of energy-rich underground tubers. Cooked food, being easier to digest than raw, increased the available energy supply and freed us up to hunt. If so, then the cooking hypothesis is not an alternate to the hunting hypothesis. Rather, it is one complementary to it; both cooking and meat eating would have promoted reduction of gut size, improved speed, and range of movement, and permitted even more hunting.

At this point in the argument, there will surely be skeptics who will doubt that our ancient hominid ancestors could become specialized enough as endurance predators to outrun swift prey that had *already* evolved to outsprint the world's swiftest predators. To elaborate the hypothesis, it is now necessary first to examine what some present-day apes do routinely, and then to consider the evolution of our uniquely human physiology, social structure, and psychology.

Chimps are generally considered to be frugivores. Nevertheless, they prize meat and hunt monkeys, young antelopes, and other mammals. Studying hunting parties of male chimps at Gombe in 1995, Craig Stanford found them depleting one-fifth of their prey population of colobus monkeys each year. Group hunts are effective. In just a half day of observing olive baboons in Kenya's Amboseli Park, I once saw a troupe of about fifty individuals catch a hare, tear it apart, and eat it with great gusto. Only two or three individuals of the dispersed troupe chased the hare, but the panicked animal was intercepted by others.

Hunting by chimps and baboons is largely secondary to other foraging, but prey is almost routinely taken as opportunity affords. These primates do not rely on meat, nor do they travel long distances in pursuit, yet they eat meat when they can, and sometimes they even hunt systematically and with great vigor. In short, even hominids that are unspecialized for a meat diet are willing and able not only to eat meat but also to hunt.

If our hominid ancestors millions of years ago on the hot open African grass plains *relied* on meat, they would presumably have evolved physiological adaptations to help them obtain it. Animals evolve unique features and capacities when they face unique situations. Of all the insects, for example, only the Apache desert cicada, *Diceroprocta apacha,* has evolved a sweating response. This insect has water available for use because it sucks plant juice, and sweating not only permits it to be active at noon on the hottest days in the hottest season of the year, it also chooses that time to be active, when it escapes its avian predators, because they are forced to yield the field. Similarly, the uniquely heat-tolerant Saharan desert ants, *Cataglyphis bambycina,* become active only when their major predators (lizards) retire from the heat. A similar scenario of taking to the field when the great predators were forced to lie low, and were less able to vigorously defend their kills, likely applies to our hominid ancestors. Humans are unique, as I will show, in having an ample sweating response that allows them to engage in sustained running in the heat, even under direct sun. Furthermore, our 3 million sweat glands excrete not only water for cooling but also toxic metabolic wastes, such as the ammonia and uric acid that are produced when we eat meat.

During my year on the bird-hunting expedition of Tanganyika, I experienced what ancient hunters were up against. I shall never forget my feelings of dreary claustrophobia during the months we spent in wet, dense, dripping mountain forests. These times contrasted with feelings of glorious exhilaration when I was out on the open savanna steppe with its scattered acacia trees and large vistas. On the savanna, to catch even small birds, I had to wander extensively. I wandered half of each day until it was time to return to camp, where my mother cooked our meals and prepared the day's specimens. I never carried water with me, to avoid being encumbered, but I was often forced to slow down or rest due to the heat. Although the heat often made my bird hunting difficult at midday, I could still travel freely. I could cope because I sweat copiously.

Internally generated heat in an animal forced to keep moving or exercising in the heat on the open plains under the African equatorial sun is one of the most potent factors limiting endurance. I could literally feel that fact, and I had shown experimentally that hawk moths, even without solar heat input, are limited to about two minutes of exercise even at modest room temperatures if their mechanism for getting rid of metabolic heat is disrupted. Similarly, jackrabbits, kangaroos, and cheetahs, even without experimental disruption of their heat-dissipating ability, are limited to only a

Jackrabbit

few minutes of running under common field conditions. It is a reasonable assumption that our ancestors would also have experienced selective pressure not only to get rid of body heat by sweating, but also to reduce heat input from the sun in order to maintain sustained physical activity at a time when they had perhaps the most to gain.

Different animals have evolved diverse means of dealing with often debilitating direct solar radiation. In New Guinea, close to the equator as in Central Africa, I found that butterflies heat to lethal body temperatures in as little as one minute if subjected to direct sunshine and prevented from using their wings for shading.

Our erect posture (with our consequent bipedal locomotion) would have been a preadaptation for us in the equatorial sun, both reducing the total amount of direct exposure to solar input and simultaneously increasing the area of exposure of skin to moving and cooling air. The tops of our heads would have been the area where solar input is focused to, potentially endangering the extremely heat-sensitive brain because of its already high internal heat load from metabolism. Thus, although bipedalism reduced overall heat input, it would have accentuated local heating of the most heat-sensitive body part.

A solution to this problem evolved. The human brain has a special network of veins that acts as a heat radiator to dissipate the extra heat load. Vein tracks on fossil skull bones indicate that the gracile australopithecines already had the same blood circulating network; this indicates that they had experienced strong selective pressure to prevent overheating.

Insects have analogous solutions. I heated honeybees on the head and discovered they not only regurgitated liquid for cooling, they also pumped more blood through the head to help carry the heat away. As in other animals subjected to potential overheating from the sun, upright hominids on the open equatorial plains likely would have evolved heat shields to reduce the solar heat input to the brain. Desert ground squirrels shield themselves with their bushy tail, desert beetles use their wing covers, camels have humps and thick dorsal hair—and we are unique in having bushy head hair that covers both head and shoulders from the sun's rays. Head hair probably evolved in part for the very purpose that it now can still serve, although it later could also have become a sexually selected trait. Later still it would also have served as insulation to reduce body heat *loss,* after *Homo erectus* left Africa and invaded the mammoth steppes of the north and became ever more reliant on a meat diet. That latter invasion occurred recently, only about sixty thousand years ago, and it may have coincided with the inventions of spear throwing and clothing.

Our nakedness and exceptionally numerous and well-developed sweat glands are potent features that contribute to running speed under external and internal heat. Because of sweating, we can tolerate very high heat loads derived from internal metabolism and the exterior environment. But the endurance that capacity buys costs us much water. On a continuous run of over 60 miles on a moderate to cool day, an ultramarathoner may lose nearly 20 pounds of water by sweating alone. Without sweating, running speed and range would be dramatically reduced. Most arid-land animals are compromised in endurance because they are highly adapted to conserve water. The fact that we, as savanna-adapted animals, have such a hypertrophied sweating response implies that if we are naturally so profligate with water, it can only be because of some very big advantage. The most likely advantage was that it permitted us to perform prolonged exercise in the heat. We don't need a sweating response to outrun predators, because that requires relatively short, fast sprinting, where accumulating a heat load is, like a lactic acid load, acceptable. What we do need sweating for is to *sustain* running in the heat of the day—the time when most predators retire into the shade.

Our ancient legacy as endurance predators is now, in "Western" cultures, effectively masked by recent changes in our ways of living. The Khoisan people of Southern Africa (Hottentots and Bushmen) were well known for being able to run down swift prey, including steenboks, gemsboks, wildebeests, and zebras, *provided* they could hunt in the heat of the day. The Tarahumara Indians of northern Mexico chase down deer till the animals are exhausted, then throttle them to death by hand. The Paiutes and Navajos were reported to do the same with pronghorn antelopes. Australian Aborigines chase down kangaroos, but only by forcing them to reach lethal body temperatures.

Each predator capitalizes on its strengths, brought to bear on the prey's weakness. Most predators catch their prey by a combination of surprise and sprint, or by singling out the young, old, or weak. In turn, prey escape by sprinting. Since they usually will not be pursued for very long, it pays prey animals to sprint fast, a behavioral trait that the human predator can exploit. As the previously mentioned anecdotes from my friend Barre Toelken have suggested, chased deer have little sense of pace. The sprints cost them dearly in the end. If the predator is not induced to give up after seeing the deer's brilliant sprint exhibition, then the accumulated lactic acid and body heat can be exploited. Humans who capitalize on the deer's weaknesses by having a longer vision—a view further into the future—can be a superpredator through the agency of mind power.

It is a truism that animals have evolved to match their morphology and physiology, along with their behavior, creating a coherent unit that fits them to their environment. In Africa, one can distinguish the European migrant birds from the residents by their longer, narrower wings, which in turn indicate greater flight endurance and their behavioral responses of launching themselves biannually on migration. Owls have eyes and ears tuned to detect mice, the unique behavior of hunting by sitting still at night and then pouncing to grasp prey with their feet, and hooked bills for tearing flesh. Kingfishers have sharp, long bills for catching fish, the physiology to digest fish and be nourished by fish proteins, and perhaps more important, the very specific behavior of diving down from a perch at moving objects underwater.

Our behavioral and psychological tendencies are also matched to the structure of our bodies, to adapt us to the environment we faced in the past.

The early pack-hunting hominids likely would have been at least as flexible in their hunting behavior as packs of African wild dogs and wolves are today. Specialized skills are required to kill zebras and bison, and learned skills in these canid groups are handed down over the generations. The more learning, the more possible diversity, so we can't draw absolute conclusions.

We are behaviorally much more flexible than most other animals. Proximally we are now so flexible that we get food in any way that we have to; that obscures our innate tendencies. We probably don't work on an assembly line or as a bank teller because that is what we prefer doing above all else in the world. We may not really know what might suit us most of all, because we don't get the exposure to find out and we become culturally biased. I had the opportunity to be out in free, wild nature to hunt. Of course, I no longer shoot little birds with a shotgun, though I still marvel at the excitement I used to feel, and that ornithologists almost universally felt at the turn of the century when they discovered new birds.

I grew up in Maine hunting deer, and that seemed to me the most absorbing activity humanly possible. I still participate in the Maine deer hunt in the fall. Getting my meat from an animal that is wild and has a chance to escape, rather than from one confined in a factory pen and raised for slaughter only, is just part of the reason. Aside from moral considerations, I hunt because of the allure. I wander the woods for days, searching for clues, hoping to see signs, and getting excited by every track. But I'm rarely "successful." Every fall, I hope that I'll get that big buck, but it eludes me. Why do so many of us bother to hunt when the chances of success are so slim? The answer came to me on a recent trip to Yellowstone Park. I saw elk, bison, bighorn sheep, and mule deer from within several yards. As I saw these beautiful animals that were tame, I knew that even if hunting them were allowed, I would have not the slightest desire to do so. The very idea was repulsive. Why? Because simply *shooting* animals is not *hunting* at all. Not even close.

It is not killing that motivates, nor is it the prize as such. The allure is in being out in the woods, in having all senses on edge, and in the chase. The white-tailed deer in the Maine woods are alert to scent, sound, and sight. They are shy, swift, wily.

The qualities that attract us to hunting are precisely the ones that dissuade the other great predators, those that do not have to chase their prey very far. Those great cats and hunting dogs take not that which is most difficult, but that which is easiest—they are very selective, trying to take the old, the young, the weak, the diseased; and the most preferred of all are the already deceased.

We are a different sort of predator. We can't outsprint most prey. We are psychologically evolved to pursue long-range goals, because through millions of years that is what we on average *had* to do in order to eat. To us, even an old deer that had not yet been caught would have required a very long chase. It would have required strategy, knowledge, and persistence. Those hominids who didn't have the taste for the *long* hunt, as such, perhaps for its own sake, would very seldom have been successful. They left fewer descendants.

Our ancient type of hunting—where we were superior relative to other predators—required us to maintain long-term vision that both rewarded us by the chase itself and that held the prize in our imagination even when it was out of sight, smell, and hearing. It was not just sweat glands that made us premier endurance predators. It was also our minds fueled by passion. Our enthusiasm for the chase had to be like the migratory birds' passion to fly off on their great journeys, as if propelled by dreams.

A quick pounce-and-kill requires no dream. Dreams are the beacons that carry us far ahead into the hunt, into the future, and into a marathon. We can visualize far ahead. We see our quarry even as it recedes over the hills and into the mists. It is still in our mind's eye, still a target, and imagination becomes the main motivator. It is the *pull* that allows us to reach into the future, whether it is to kill a mammoth or an antelope, or to write a book, or to achieve record time in a race. Other things being equal, those hunters who had the most love of nature would be the ones who sought out all its allures. They were the ones who persisted the longest on the trail. They derived pleasure from being out, exploring, and traveling afar. When they felt fatigue and pain, they did not stop, because their dream carried them still forward. They were our ancestors.

Sometimes I wonder if this ability to have long-range vision, if not also the drive to explore, might not also have been the boost that gave us our *unique* brain power to extrapolate. The currently popular explanation of our unique intelligence is thought to be related to *deception* in the social context. Deception indeed tweaks capacities for mental visualization, and there is little debate that social interactions involve keeping track of individuals, trading favors, paybacks, and possibly deceiving. In support of this idea of intelligence based on sociality, brain size in animals correlates with group size. Did, then, *Homo erectus* live in superlarge groups relative to other animals? That's unlikely; these archaic humans were hunters, likely living in small groups, and their brain size already overlapped with that of moderns. Another hypothesis, also a nonexclusive one, is that sexual selection was a driving force in the hunting syndrome. There is no either-or answer; all factors likely acted in concert, but I'll briefly examine the last hypothesis.

The mere fact that we can run down some of the swiftest ungulates, animals that have evolved to outrun the swiftest predators, indicates that we are indeed highly specialized, physiologically and psychologically, for that particular task. But there is a sexual difference. Curiously, in all human cultures that have been examined, as well as in baboons and chimpanzees, hunting is largely a male activity. Sexual specialization is common in animals. In some hawks, for example, females are larger than males and catch the larger prey, while the males specialize in the smaller prey. As a consequence, the sexes in effect achieve a division of labor in foraging, causing less depletion of the food supply near the nest.

For the protohominid females, pregnant or burdened by offspring who needed to be carried along, hunting for large animal prey requiring long pursuit was even more difficult than it is for present-day apes. Having become naked to increase heat loss, the young could no longer hang on to their mothers' fur but rather had to be held. Through food sharing, a multifaceted male-female symbiosis evolved. Adult men were free to hunt, but women foraged and chose mates. On what basis did they choose?

The females with young could not readily take part in long exhaustive hunts, and they needed to enlist the aid of males to provide them and their families food. Hunters killing large animals had a temporary superabundance of meat, which could not be stored. How could it be used? It was brought home, to be shared in mutual obligations with other hunters, and to trade for sex. Sleeping together and eating together became interrelated. It is an old formula. Chimps trade sex for food routinely, as do baboons. Craig Stanford, who studied the hunting practices of chimps at Gombe, says, "Chimps use meat not only for nutrition; they also share it with their allies, withhold it from their rivals. Meat is thus a social, political, and even a reproductive tool." Similarly, among the Aché of South America, women prefer successful hunters, the meat providers. Similar correlations between reproductive choices and resources exist in most societies where the limiting factor to reproduction in females with young is resources. For males, the limitation is more commonly sex.

For !Kung Bushmen, meat is only a small part of the diet, but it is the food they desire most. The women bring in the bulk of the food, feeding the band from day to day on berries, bulbs, leaves, and roots. The men hunt, oftener than not having little to show at the end of the day. Still, hunting is deemed highly important. Only after a boy makes his first kill of a large antelope does his father perform the rite of the first kill, which marks his passage from adolescence to manhood. A male !Kung who does not hunt remains a child who cannot marry. He cannot expect to have a wife if he cannot bring home meat and skins for his family and parents-in-law. Bushmen males hunt from the time of adolescence until they are old. Often they travel 30 kilometers per day, come home with nothing, and by next morning are off again, impelled by their will to persist, if not their wives' goading. They carry no food or water with them, because that hinders their ability to travel. !Kung hunters might follow a wounded giraffe for five days. This is not work that women carrying infants can perform. Division of labor, though perhaps currently not politically correct, is an ancient tradition with deep biological roots. And there is nothing wrong with diversity, either between people or between sexes. Division of labor has allowed men to rely on women to feed them, and enabled them to engage in long-range hunts after large prey that required traveling rapidly and unencumbered over long distances.

Contrary to some presumptions and misconceptions, the idea of *man the hunter* as a driving force in human evolution neither denigrates women nor relegates them to a passive role. Misconceptions can be minimized if we read "man" as "humankind." Evolution has not very likely affixed the huge complement of genes that affect growth and development of the brain and human evolution onto the Y (male) chromosome. The different behavioral tendencies of men and women, at least those regarding long-term cooperation to rear children, can best be explained in terms of compromise and cooperation. If "man" is the hunter, then it is because women permitted or selected him to be. They are the other half of the same man-the-hunter syndrome. Women had to become *intelligent* choosers, because choice could not be trusted to appearance alone.

Sexual selection in the animal world often results in such runaway scenarios as the notorious peacock's tail, the elaborate songs of some birds, and even balloon making in flies, which I will discuss later. If the hypothesis that hunting is, in part, also a consequence of sexual selection, rather than merely a device for the fulfillment of caloric requirements, then there are wide repercussions, because with energetics removed as the primary constraint on hunting, there are few limits. If bringing a rabbit back to the group can enhance a hunter's sexual desirability, then just imagine if he kills a mammoth or has the ability to do so, while the woman develops the capacity to evaluate!

The strategy of supplying protein to secure mating privileges is the rule in many male birds, spiders, and insects, especially in scorpion flies (*Panorpa*), some grasshoppers, crickets, and cockroaches, and some beetles (Malochiidae). Among insects, the nuptial gift food offering may be prey, or in the absence of prey, protein from body secretions. In some mantids and spiders, it is the male's own flesh.

Male mantids are legendary suitors, who regularly make the ultimate sacrifice for sex: their own bodies. They are cannibalized by the females with whom they mate. The benefits of offering themselves (usually reluctantly) may go beyond just providing a dietary supplement that ultimately provides nutrition to the eggs he has fertilized. The females first eat the male's head. Males with intact heads mate for only four hours, but the encounters that result in decapitation last up to twenty-four hours. Maydianne Andrade, who studied Australian redback spiders, has shown that being eaten prolongs copulation and increases the amount of sperm transfer. In this spider, allowing himself to be eaten also prevents the female from mating with other males, since a satiated female spider rejects other suitors and the "victim's" sperm thus have precedence in fertilization. Spiders die for sex, and their suicidal behavior has, ironically, evolved because it increases their individual fitness.

These extremes alert us to mechanisms that might otherwise remain hidden in our species. All animals have to pay something for sex. My favorite example, because it is in some ways a caricature of the human situation while at the same time illuminating of the evolutionary process, is that of dance flies (Empididae). These European and North

American flies are predators that hunt other flies. Their name is derived from their group gatherings, during which these insects fly up and down and sometimes in distinctive lines and curves, or dances. Females choose mates from among the dancers on the basis of the male's energetic displays and the male's offerings during those dances.

To have a chance of mating, a male must sustain himself in hovering flight while holding a nuptial offering of a fly carcass in his feet. Females size up the lineup, then pick and choose. Couples then drop to the ground, where the males transfer their offering to the females, which then eat their offered prey.

A small fly carcass suffices as a suitable mating inducement in some species of empids. Another step in the evolutionary progression is found in other species, where the male wraps the prey in a fine shiny veil that he weaves with spinning glands on his forelegs. A silk-wrapped nuptial gift is more attractive to the females than an unwrapped one, possibly because it appears larger in size and is more conspicuous than unadorned or unadvertised prey.

The next step in the evolutionary progression seems downright devious. Some males dance carrying an even larger and more conspicuous package, but one that contains a fly that is too little to eat (but easier to carry and thus show off), an inedible piece of debris, or nothing at all. For example, in *Empis politea,* the male carries a great white egg-shaped balloon within which he may or may not enclose a little fly. The female pays no attention to a male carrying a fly. She goes for a big, showy but empty package.

In *Hilaria sartar, S. sartrix,* and *H. granditarsis* the males have taken the final step in deception. They always carry an empty gleaming white oval balloon throughout the course of their ever more acrobatic dance. If a foolish male were to attempt to hoist a balloon containing prey, he would be badly outclassed by dancers with lightweight balloons. Hoisting heavy prey offering would be possible only at high temperatures, when the fly's muscles could achieve the high work output required.

The flies prove the point that, as with the scenario that anthropologists have proposed for humans, it is not the nutritional value of the offering that counts. The showing off does. One difference is that human hunters can't cheat. In the protohominids, the males had to bring home real meat, or demonstrate ability to do so, not just hoist pretty empty packages. Meat was a valued resource that was an important and necessary part of the diet.

As in the dancing flies, vigor or capability can sometimes be evaluated on the basis of physical appearance, but it is more reliably based on *performance,* either in hunting or in symbolic representation. Could the origin of our dances, like our athletic games that provide worthless colored ribbons and metal or fake metal trophies, be symbolic activities that show off our capacities?

A race is like a chase. Finishing a marathon, setting a record, making a scientific discovery, creating a great work of art—all, I believe, are substitute chases we submit to that require, and exhibit, the psychological tools of an endurance predator, both to do and to evaluate. When fifty thousand people line up to race a marathon, or two dozen high schoolers toe the line for a cross-country race, they are enacting a symbolic communal hunt, to be first at the kill, or at least to take part in it.

The real hunt is long out of date to most of us. Very recently (geologically) we eradicated some of the most magnificent creatures that were ever on this globe when we came, as full-fledged hunters, into contact with them in

America, Australia, Madagascar . . . We had by then evolved the psychology, physiology, and technology to make us extraordinary hunters. In contrast to those in Africa, our homeland, our new prey did not have time to evolve effective evasive responses to our unique hunting capabilities, which combined physiology and psychology with intelligence, and ultimately also weapons.

It is fortunate that we have now invented some ecologically friendly redirections of our hunting tendencies. We can now chase one another rather than mammoths and mastodons. We can be road warriors, who will have races to run forever. Now we dream not of killing great beasts to be heroes as we provide nourishment for our social band. We may dream—and get the same psychic nourishment that was once necessary to provide bodily nourishment—of winning races or setting records or fulfilling other long-term goals. In the Olympics we witness the biggest hunt. If we can't be part of it, then as spectators we cheer for those who represent us, who are really a part of us, since through our evolutionary time of millions of years we were (and still are) mutually interdependent. There is one main difference, though. In contrast to hunting prey animals, where there is always an end point, in chasing against one another there is no end point. Where can it end? What are the limits?

FOURTEEN

Running Like Dogs and Cats

Cheetahs and wolves represent two extremes of hunting strategies. Both employ running, but different kinds of running and in very different contexts. Felines hunt solitarily for the most part, and their success depends on being able to spend much of their time in waiting. Using stealth, they succeed in getting close to prey or vice versa, then they explode in a lightning-fast sprint. In contrast, the canine strategy generally is to work in teams and to identify and then chase prey with weaknesses. Canines choose carefully and can pursue over a longer distance. Specific muscle types match psychological tendencies. Cats have a preponderance of fast-twitch fibers, which rely on glycolysis of carbohydrate for explosive release energy. Dogs have more slow-twitch fibers, which require oxydative metabolism using fats. In us, both fiber types are used during sustained performance, although elite sprinters have

Cat

about 26 percent slow-twitch fibers, versus 79–90 percent for elite distance runners. Fiber physiology is, however, not entirely fixed but alters with training. For example, during endurance training the lactate produced during glycolysis by the fast-twitched fibers accumulates less because it is more rapidly cleared.

There is leeway in behavioral tendencies between species as well. For example, a cougar relies mostly on stealth and surprise and may chase a deer for only a few seconds. On the other hand, a cheetah on the African plains may be forced to chase an antelope for up to a minute or more, either until catching it or until having to stop from heat prostration. The much slower lions often work in groups. However, the main feline and canine tendencies are still apparent in our house pets even after centuries of domestication. A house cat will seldom follow its owner through woods and fields to hunt rabbits as a dog would with its companions, in this case us, with most eager anticipation.

Dogs will run even when they are not hungry, and they

derive pleasure from the hunt itself. Dogs will gladly retrieve such symbolic prey as sticks and Frisbees. The pleasure they get from these activities is, as running is with us, partly social. What human runner would race with the passion of a dog retrieving a thrown stick, if it were not for the incentives thrown out by others? Without others' interest the time to finish a race or run a specific distance is a dead and uninteresting number. Cats are not socially motivated like dogs or humans. No matter how many times you make a cat run around the track with others, it will still not race, nor can it run as far as a dog.

Dogs are the descendants of wolves, and since they can interbreed with them to produce fertile offspring, they essentially are still wolves. It may take a stretch of the imagination to recognize a manicured poodle being led on a leash down a city street as a close relative of the pack-hunting wolf capable of pursuing and killing moose. However, like his equally domesticated human counterpart, the poodle still has the wolf in him. His pack loyalty has been transferred to a master, and he strains at his leash to go out for his daily romp. He also still likes to eat meat rather than grass.

One might think that the original wild-type wolves might be the choice runners for the dog's ultimate endurance running event, the Alaskan Iditarod race. They aren't. No wolf races like a husky, because running capacity as such is not enough. Many types of dogs have been tried for racing. For example, Johnny Allen, an Iditarod winner, has dominated Alaska dog races with his mix of wolves, huskies, and Irish setters. No matter what it may look like, a dog is still a dog and running distance is a general dog talent. But for Iditarod racing, the runners have to feel part of a team, a pack.

Iditarod dogs require a strong appetite, a deep chest, and

Canine

a strong cardiovascular system, which wolves and many dogs have. These are not the limiting racing factors. What is most strongly selected for in any line of dogs destined to be Iditarod ultraracers is the desire to run. Those destined to race are the dogs that seem overwhelmed with eagerness as they strain to take off and to keep going, and going, and going. There is no apparent difference in either athletic capability or intelligence between racers and nonracers, even though the racing capacity is the result of an ancient heritage of endurance-running predatory ancestors. Many types of dogs can be turned into racers by proper training and environment, but that's not enough to make a racer out of a cat.

Perhaps I had ingrained husky characteristics, being especially talented in the requirement of having a good appetite. In my high school yearbook it was noted, "Ben likes to eat and run." I ran to school then, and I run to the office from my car in the morning now, simply because I don't like walking. Running feels good, and it saves time.

It is not always easy to tell what physiological factor is relevant to being able to run fast or far, because there are so many variables, and simple desire may swamp them all. As soon as I thought I was inherently flawed because my hero was tall, I'd find another, perhaps even better runner who was short. I'd then think you had to be very skinny, and then the next hero to come along would be muscular.

When I was in college, I thought you had to be white and from northern Europe, preferably Scandinavia or Ireland, to be a world class distance runner. The distance events had always been dominated by northern Europeans, immortalized by Paavo Nurmi, "the flying Finn," and the Czech Emil Zatopek, but a host of up-and-coming British and Australian runners were beginning to decisively dominate the middle-distance events. One almost never saw a black long-distance runner. Conversely, many of the best sprinters were black. We all thought that sprinters are born but middle- and long-distance runners are made. It was assumed as an empirical, scientific fact that blacks are innately blessed with natural speed, so that they have raw talent for explosive events. The correlating assumption was that blacks lack either the stamina beyond a quarter mile or the dedication to make themselves into middle or distance runners. Then, ironically, the "evidence" shifted 180 degrees, and it seemed you had to be black and from equatorial Kenya, Tanzania, or Ethiopia to be world class distance-running material. Already, in the Rome Olympics of 1960, Abebe Bikila from Ethiopia ran barefoot into the stadium, winning first place in the marathon. The real surprise came in the 1968 Mexico City Olympics, when the Nandi tribesman Kipchoge (Kip) Keino from the small African country of Kenya beat the until-then-invincible greatest American middle-distance runner, Jim Ryun, in the mile.

Keino was the most spectacular Kenyan running success in the Mexico Olympics, but eleven other Kenyan track athletes also won medals there. Since then, the Kenyans have dominated long-distance running events. Black Kenyans are winning everything from 800 meters to the marathon. In 1999 the Olympic men's marathon A standard was reached by 240 runners worldwide, and seventy-six of them were Kenyan, which was almost twice as many individuals as for any other nation.* New African runners are popping up almost every month, each as good as or better than the next, and now almost all marathon finishing times would have left the flying Finn and all of his northern European compatriots far in the dust—indeed, by my calculations, not yards but close to 3 miles behind! Clearly, the preconception that blacks lack endurance has been turned on its head. A new presumption is that Africa's high plains are some kind of special sieve that has selected for the swiftest, wildest, strongest beasts and men, so that these Africans are genetically predisposed to succeed at all running events!

Another explanation that was offered relates to elevation. The Mexican Olympics were held at seventy-three hundred feet, and the Kenyans live at high elevations as well. There is less oxygen available from the thinner air at high elevations, and our $\dot{V}_{O2}$ max decreases by about 3.2 percent for every thousand feet in elevation above five thousand, in those of us not accustomed to altitude. One hypothesis to account for the Kenyans' running success is that living at altitudes of over a mile has adapted their respiratory systems to give them an advantage at altitude. With time at high elevations, the body adjusts to bring the $\dot{V}_{O2}$ max back up. However, sea level performance is not increased by training at altitudes, and the altitude hypothesis can thus not account for Kenyans' superior performance, still exhibited when they compete at sea level.

*As this was going to press, Ethiopians took first and third place in the Sydney Olympics Marathon, and a Kenyan was second.

According to a recent book by John Bale and Joe Sang, the real explanation for the Kenyan running phenomenon is cultural. These authors point out that most Kenyan runners come from one locality. There are distinct running regions in Kenya, and by far the highest per capita running medals captured on the world running circuit belong to one group, the Kalenjin, and even more specifically to one group within the Kalenjin, the Nandi. It is unlikely that of all the African blacks, only the Nandi would be blessed specifically with physical adaptations that translate to running prowess. There must be something else. According to data compiled by Dirk Berg-Schosser, in contrast to other Kenyan ethnic groups, this tribe has the highest "achievement orientation." They are generally regarded as quiet, ascetic, serious people who are hardworking individualists. A principal at a school at Kapsabet even said that the Nandi were "too much individuals to learn to play football or hockey as a team. They usually prevent themselves from being beaten too badly by a good team by sheer guts."

The Nandi's traditions derive from their culture as cattle raiders (their "sport"). Athletic prowess came to be prized because of traditions. When one of their tribe, Kipchoge Keino, stepped prominently onto the world stage, they suddenly had a role model. To them, racing became the only game in town, as it had been before to the economically deprived Irish and Scandinavians, and on a micro scale to those of us at Good Will. Running is perhaps the most fundamental of all sports, and it is economically the least costly to perform. As a consequence, it is the most democratic and most competitive of all sports because individual merit can prevail despite economic inequality. It is a sport for everyone, the whole world over.

Running, throwing, jumping, the repertoire of track events, are the basic body movements required for hunting and warfare, and they have been ritualized into games, dances, and initiation ceremonies. It is a small step from there to racing. To the Nandi, racing has replaced former activities that required the same individualism, fortitude, discipline, hard work, and ability to delay gratification. In former times the Nandi aspired to be *barngétung,* a name given to a member of the tribe who had succeeded in spearing a lion. Similarly, their success at cattle raiding from distant neighbors depended on days of trekking through hot arid country, which relied on physical stamina, hard work, sacrifice, and ability to endure hardship. It was exhilarating sport, and it gave them something to live for. Racing became a ready substitute to warring and raiding, as it can be for hunting large animals.

Similarly, the marathon is also a cultural offshoot derived from warfare. This race of 26 miles 385 yards is a ritualization of a specific historical event. The name of the race derives from the name of the village in Greece from where, according to legend, Pheidippides, an Athenian professional military runner, ran the now standard marathon distance to Athens to report the happy news of victory over the Persians in a decisive battle fought in 490 B.C. Reportedly he said, "We conquer," then fell over dead.

Within the Nandi, Kip Keino became a hero and a symbol of manhood. He founded a tradition, as Pheidippides did for the Greeks and later Western civilization. Like Pheidippides, Keino came home to report symbolically that they had and could conquer giants. There were then few outlets for Nandi to display prominence, since lion hunting, cattle raiding, and warfare had been outlawed. Here was a new chance for glory through a redirection of old traditions.

At age fifteen Keino ran a mile in only 5:56. Training full-time as a government employee, he worked extremely hard and incorporated all of the latest training methods.

Subsequently, talented individuals were recruited to the high schools at which organized training and competitions were established. The selection of talent by the state and its nurturing was important in producing the Kenyan running phenomenon.

Each successful Kenyan racer who makes the world scene is directly or indirectly selected from the whole population. The population's average is irrelevant for Olympic performance. What matters is that the population contains at least a few potential stars. It may well be that on average there are more *potential* basketball stars among a given number of Watusi or Masai than among the same number of residents of Boston, but that says little about where the gold medalists in that sport will come from. It may well be that most black women in East Africa walk while balancing huge loads on their heads as if it were a natural thing to do. Sure, it's genetic. But nobody could rationally propose that European women are not equally genetically endowed to be able to do precisely the same. They just don't have the same exposure to a culture that fosters the practice from very early childhood and then channels development. Why should it be different regarding racing success, where only a vanishingly small percentage of the population (not the mean or average) gets to achieve prominence? Talent is not enough.

Bale and Sang conclude: "In athletics it is culture and not biology, attitude and not altitude, nurture and not nature, which are crucial variables which 'explain' individual athletic success in the rationalized and regulated world of achievement sports." Tom Derderian, who wrote the book on the Boston Marathon, concurs:

> The idea of talent is a myth because talent is only manifest in retrospect. The myth should be smashed,

> toppled, because it is an insult to consider an athlete who made it to the top, unfairly and inevitably because of his or her genes, and not because the athlete decided by thinking in an act of free will to try to win. In that intelligent decision, an athlete engaged the competition with a pledge to win. With that intention, support of community, knowledge, understanding, and the willingness to take a risk and suffer the consequences comes no guarantee, but just the possibility of greatness. The talent lies not in our genes but in our minds.

Few people know what goes into a championship performance unless they've tried it themselves. As Bill Rodgers, winner of four Boston and four New York City Marathons, said, "Running is the world's greatest sport, with the best athletes. It requires so much physical and mental energy, a real commitment." However, one truth does not necessarily exclude another. One cannot succeed in racing unless one has the health and physique, that is, genes. We're not all identical peas from the same pod who happen to roll in different directions. There *are* individual differences, and for one individual it might be relatively easy to train to run a mile in under 5 minutes or a marathon in under 3 hours, while to another individual such feats are heroic. Heroism deserves credit wherever it is exhibited. To run a championship performance requires commitment and willingness to take risks, but the risks are greater when the talent is less. Those who seemingly show less greatness because they run in the pack may actually draw from a fount of courage. How much can gallantry overcome? Before setting up a specific running goal, one must be realistic; it is important to know if you are a cat or a dog.

When Ed Styrna, my college running coach and a former national champion in the hammer and javelin throws, was asked (in retrospect) to comment on my running, he wrote, "Ben would have been left in a cloud of dust after the starting gun went off in a 200 meter race against Michael Johnson, but Johnson would have been foolish to try to keep up with Ben in an ultrarunning race. Ben had the kind of muscles that were designed for endurance and he had the mental toughness to back them up."

Athletic achievement is contingent on a realistic assessment of one's capabilities. Nobody will consciously focus his life for years on end, to the point of enduring agonizing strain and discipline up to the breaking point for possible victory and glory, when victory is not attainable, or when glory will be snatched away even if victory might be attained. When the bar is impossibly high, few will aspire, and when there is no glory, few will be inspired. Ultimately we are shaped by the anvil of our environment and the hammer of our mind, if used.

The dream is important. In us it activates the brain, and in all creatures, even insects, the brain activates the body. Insects presumably don't dream, but in some species subtle environmental cues activate the nervous system to flood the body with hormones that cause massive muscle growth as well as engendering all sorts of other changes. In some aphids, mere changes in day length result in them growing wings and the musculature required to power them. Surges of hormone production resulting in huge physiological changes of the body are universal not only for insects, but also for reptiles, amphibians, birds, and mammals. If some animals' brain hormone production can be triggered by mere flashes of light and other numerous and seemingly trivial external cues, then it does not seem preposterous to consider that just maybe we can be molded by fierce dreams that allow us to perform what we'd otherwise be incapable of accomplishing.

My commitment to win the 100-kilometer National Championship involved a large risk, because I could be reaching for more than I was capable of. I had to remodel myself within the constraints of my "normal" physiology to be able to do what would not otherwise be possible. But what *was* possible? When I had won the Golden Gate Marathon, been first master at the Boston Marathon, and been third in my first 50-kilometer race, the deciding moments had been near the very end. Perhaps I had discovered my strength. To not use it fully to try for an inspiring goal seemed wasteful, if not disrespectful, like foolishly squandering a precious gift.

FIFTEEN

The Fitness of Being Out of Shape

Nature has neither kernel nor shell; she is everything at once.

—GOETHE

While studying the temperature-dependent endurance physiology of hawk moths at UCLA, I raised hundreds of these animals from caterpillars, feeding them fresh tobacco plants that I grew in the greenhouse. The big green larvae, plump and slow creatures that they were, molted into seemingly comatose, limbless brown pupae that lay motionless like mummies for about two weeks. After the two-week interim, each animal was totally transformed into a moth with wings, legs, and a tubular tongue as long as its body, curled up neatly into a roll under its head. The moth slipped out of its pupal shell, crawled a few inches up onto a twig, and dangled there limply for a couple of hours while it gradually expanded and hardened its wings and the rest of its exoskeleton. By evening the new moth would begin to shiver a minute or two to get its huge set of powerful muscles hot enough to contract rapidly and strongly enough for flight, then lift off on whirring wings to execute a superbly coordinated behavior. At the end of warm-up and during flight, these muscles contracted at 40 times per second, requiring an energy expenditure with the herculean aerobic rate of about four times the $\dot{V}_{O2}$ max of a pronghorn antelope running all out. Incredibly, this high rate of aerobic metabolism appeared with essentially no training (some muscle contraction in the pupal stage is possible), and as far as we know, exercise does not increase the moth's already high aerobic metabolic rate. Similarly, many birds leave the nest flying, requiring very little if any training or conditioning, but maturation instead.

It seems intuitively obvious that being instantly in shape, as these moths and many birds are, might be advantageous in short-lived animals that do not have years or months to get ready for their life tasks. These hawk moths feed on nectar and can live for several weeks. Some other hawk moths, who don't feed at all, live only a few days as adults. Saturniid moths are born even without mouth parts and they starve as their onboard energy reserves run out after a couple of nights of flight activity; they can't waste time and energy in training.

Whenever I contemplated the difficult and often excruciating task of getting in shape for a race, I thought of the moths. Why do we get out of shape? Why is it easy for the moths and difficult for us to become paragons of aerobic fitness, in terms of both power output and motor coordination? I knew runners who could run faster and farther than I could with little training. Did I lack talent? Was my natural state to be clumsy and out of shape?

In our evolutionary history, we may have been forced to be nearly continually active in order to survive so that, in contrast to the moths, we may never have had the necessity to deal with the effects of prolonged idleness. That idea is supported by data from bones and bears: our bones become brittle and weak if they don't receive normal everyday stress such as from walking. Bears, who lay comatose for six months at a time in hibernation, suffer no such bone deterioration from inactivity. Similarly, if running had been a constant in our lives throughout evolutionary history, it may now be required as a supplement for optimum health, analogous to vitamins, which provide chemicals that our bodies have not evolved to produce because they have always been present in the food of our normal diet. We need to take exercise, and vitamins, when our normal life styles, and diet, are at odds with the ancestral conditions that shaped us. Alternately, or additionally, perhaps being able to be out of shape is adaptive because it permits neuromuscular flexibility. We can retrain our bodies, for example, from a lean runner to a weight lifter. Moths cannot. They are wired up and muscled up to do one specific thing extremely well, but at the cost that they can't remake themselves.

One cost of aerobic running fitness is loss of explosive muscular strength. When untrained, I normally bound up three stairs at a time, but I know I'm becoming trained for long-distance running when I can do only two at a time. The loss of explosive power that occurs with aerobic conditioning is thought to involve a conversion of fast-twitch, anaerobic muscle fibers to slow-twitch muscle fiber characteristics. That is, with aerobic training we lose sprint speed. Conversely, we can specifically train sprint speed, but it is at the cost of endurance. Similar trade-offs apply to being fat, another aspect of being out of shape.

When we need food, having the speed and mobility to chase it down is obviously advantageous. Once we have the food, it is adaptive to conserve what we have laboriously won. It may then be advantageous not only to slow down, but also to convert food to fat and to *keep* it in fat. Reducing mobility and becoming physically lazy would serve that purpose. Once most animals are fat, they do indeed reduce mobility and are likely to become even more fat if food is still available. Birds are an exception, because after they fatten up they quickly use that fat as fuel for migration or for shivering on cold nights. However, they don't get fat at any time, even with food available. They lay on fat at *specific* times for specific tasks.

Fat is the body's bank account. The currency is calories. For us, it is an insurance policy that's taken out for lean times up ahead. The best proof for the fitness of fat can be found in animals inhabiting a highly seasonal environment. In the Maine woods, the moose, porcupines, and snowshoe hares that browse year-round on the readily available buds and twigs are always lean. They have no need to carry energy stores, and they are not programmed to store calories for possible food shortages. They can stay lean, as is necessary to outrun predators. For most animals, continually available food translates to never being fat because there is then no need for fat but advantages to being thin. On the other hand, those animals, including woodchucks, bears, raccoons, and skunks, whose food supply disappears in winter become obese in the fall, when, given a chance, they feed as if there were no tomorrow. They've been programmed to feed precisely because the tomorrow they anticipate is one of lean times. Similarly, to our body, fasting is a danger signal that says, "Food getting scarce—stock up if you possibly can." Thus, I predict that when trying to lose fat, a very gradual

Prairie dog, fattened up for hibernation and carrying more food, from a photo by Stephen G. Maka

caloric adjustment, to try to sneak up on the body so it won't notice, is probably preferable to fasting.

Some animals, such as migrant birds that can double their body weight in about two weeks, lay on fat to an extent that outstrips even the best of human capacities. We don't know how they do it, but there are clues. Not surprisingly, physiological regulation of appetite is a big factor, and it involves chemical signals circulating in the blood that affect the appetite centers in the brain, and these agents are influenced by environmental cues. Bears are veritable gluttons all summer and fall, but in winter and spring they don't eat or drink, even if food is available. If bears were hungry in winter they would not sit still, and winter-active bears did not survive to pass on their genes, because they exhausted their fat reserves too early. Even after coming out of hibernation following a half-year fast, bears' appetites are initially still physiologically suppressed by their blood-borne chemical signals. The animals are thinned down considerably from nourishing themselves and, in the case of females, their cubs also. Appetite suppression in early spring is adaptive in them, because there is then generally still not much food available. The harvest months, when appetite appropriately peaks, are in late summer and fall.

Humans, like bears and birds, are obviously also capable of laying up fat stores, which suggests that our ancestors very likely experienced times when food was abundant, followed by times when food was scarce. Aside from seasonal changes of food availability, we could not predict famine or hunting success. Further, our ability to fatten at any time tells us that those who survived to be our ancestors were capable of fattening when opportunity arose. The benefits of storing fat are especially great for women, who have to sustain energy balance through pregnancy and lactation. Not surprisingly, even now women universally, on average, have more body fat than men and put it on more easily.

We can only speculate about why nature favored fatness in us despite its associated costs, but the examples of other animals give clues. Fatness would have been particularly useful when, as in hibernators, huge energy demands coincided with decreased mobility. In human history, women were either pregnant or lactating virtually continuously. The energetic demands of childbearing were tremendous. Also, women with children could not move around as easily as men. In a world of often limited resources, fatness was associated with fertility, and most likely considered sexy.

In females, the distribution of fat might have evolved to accentuate its presence, so that it could be shown off. Although a slim body might signal potential mobility and hence capacity to secure resources, a plump one signals something of even more direct reproductive value, namely potential success in childrearing. In the past, in most societies, food shortages were virtually inevitable. Now that food production and distribution systems have made nutrition predictable and reliable in many countries, fatness is no longer adaptive there, and hence it is no longer prized as a signal of success. Our natural tendencies are not necessarily measures of goodness. Science can help us recognize biological biases, and our values can then be engaged to either overcome or favor them.

It is possible that there were regional differences in costs and benefits of putting on fat, and that these differences would be reflected in the present. It has been proposed, for example, that Polynesian people of the South Pacific tend to be more genetically prone to be heavy than many others. The thousands of Oceanic Islands were colonized by a specific subgroup of survivors—people who had been adrift for months at sea. During these long-distance movements that resulted in chance colonizations, those leaving on their journeys with the largest buffer of energy reserves (like those of migrating birds) would have lived longer, and possibly more likely reached land, than those starting off lean. They then passed on their genes to their descendants. Regardless of whether the ability to be fat was later useful, it would still be passed on as a survival trait. In parts of Oceania fatness is even considered a mark of beauty.

To train for an ultramarathon, I made the rational choice to try to train my fat-burning, as opposed to fat-building, capacity to a maximum. Of course, I needed to have fat to burn, but I needed to have a very *thin* cushion of fat even though my body would naturally try to fight that tendency and put on too much.

My natural body weight—the weight I've had since high school and even now at age sixty—is right around 160 pounds, with almost no obvious fat. For my five-foot-eight height, I'm chunky for a distance runner. For example, Olympic and frequent Boston and New York City Marathon winner Bill Rodgers is the same height as I am but weighs 36 pounds less. Frank Shorter, Olympic marathon winner in 1972, is 2 inches taller than I and weighs 26 pounds less. I needed to reduce my weight. But how? Restricting my calorie intake would cause a conflict within my body. Some initial weight loss might be easy, but my body would wake up and defend what it perceived as its normal mass by restricting energy expenditure, to try to hang on to calories, which in the past were always hard won. My body would not be concerned with winning a marathon. All it knows about is long-term survival. Dieting can reduce metabolic rate by as much as 45 percent, making one weak, sluggish, and slow, while reducing weight only slightly. I needed to lose weight, but not at the cost of reducing my ability to have the high energy expenditure necessary for running fast. How could I induce my body to maintain a weight like Shorter's, rather than my usual 160 pounds, while still maintaining the capacity for a high rate of work output?

SIXTEEN

Diet

Eating is controlled by psychological drives, which in turn are influenced by blood chemistry. Just thinking about a hamburger can change your blood chemistry. A dog might actually have to see or smell the hamburger first, or at least hear a dinner bell. We can think of the effects on our bodies years ahead, and hence can influence them. Dieting alone has little effect on our brain's apparently quite strict hypothalamic body weight set point unless combined with strong will. Can the mind affect the set point? If anorexia is any indication, then our minds are indeed powerful enough to alter our hypothalamic set point. So, like an anorexic, I continually visualized thinness.

Our hypothalamus is well known to have control centers that regulate our basic requirements, including the optimum intake of calories, specific vitamins, and other nutrients. We can become dangerously overweight when the nutrients we need are in short supply in the food we eat so that we just eat more, becoming saddled with surplus calories.

In preparing myself for the 100-kilometer race, I would need many nutrients in varying amounts, and I did not presume to be able to figure out which they all were and how much of each I needed. There are no absolutes anyway. Much would depend on conditions. Nutritional requirements when running 20 miles a day in training would presumably be different from those of a normally active person. I could not figure out what I needed, but I was confident that my body would know, in the same way that any other wild animal's body knows. For example, a pronghorn antelope has a perfect running physique. It eats the right kinds of vegetation without knowing anything about diet. It simply obeys its hungers and aversions, selecting food from the diverse menu it encounters in its environment. I decided to be like such an animal. My body would choose the right things in the right amounts, provided it had the choice of a varied menu of unprocessed or minimally processed food.

There were three additions to my eat-anything-I-want dieting strategy. First, I ran a lot, thus my caloric balance was heavily biased to expenditure. Second, I visualized thinness in the same way that I visualized speed and endurance. The mental visualization itself would not affect my build directly, but it might bias my food intake and training, and thus indirectly affect the ultimate outcome. Lastly, and perhaps most important, I capitalized on my knowledge that I'm a descendant of hunting apes who probably ate meat regularly.

The nutrients that our bodies can't synthesize must come from our diet. Thus rats, who can synthesize ascorbic acid (vitamin C), have no need for it in their diet. We do need it in our diet, probably because our ancestors ate ascorbic-acid-laden food routinely enough so that there was no need to develop the synthetic pathways to make that substance.

Meat is, to us, much more than a source of calories. It contains all of the amino acids we need to build body proteins. It also contains fats essential for brain growth during childhood, iron needed for oxygen transport, the B vitamins necessary for oxidative metabolism, and vitamins A, D, E, and K. If meat contains all that an animal *is*, does it not also contain all that we need—aside from the calories burned off in exercise? This is not to say that we absolutely need meat. We can also get the same nutrients from vegetable sources or pills, but proper eating then takes much greater effort. Each animal has its own unique evolved diet. It is easy to overlook some detail when you're trying to play God. I would rely on my appetite. And the more I ran, the more I craved greasy pork chops.

Judging from later research on migratory birds, it might not have been a bad choice. Fat metabolism is biochemically coupled with protein metabolism, which is why we can't subsist solely on a pure fat diet but require protein at the same time. Migrating birds get their protein by literally consuming some of the structure of their bodies. Additionally, their and our brains use glucose and not fat for fuel, and after glycogen reserves are exhausted that glucose comes from protein degradation and hence muscle degeneration. For example, the great knot, *Calidris tenuirostris,* after flying nonstop for 5,400 kilometers between Australia and China shows weight losses of skin, salt glands, pectoral (wing) muscles, heart, liver, intestine, kidneys, spleen, and leg muscle, aside from the huge fat losses. Fasting produces similar results of body protein depletion, and I would rather get my protein out of my stomach than off my muscles and vital organs. In migrating birds and during long fasting, only brain weight remains constant, showing what is really most important and irreducible, in running and in life.

SEVENTEEN

Racing Fuel

What does not destroy me, makes me strong.

—NIETZSCHE

Robins, warblers, and most other songbirds live on an almost exclusively protein diet of worms and insects when they are young and in the nest. That is, they eat meat. Later, when they are mature and get ready for migration, they switch their diet. They require calories for energy, and they then fatten on berries and other carbohydrate-rich foods. Similarly, on their migratory stopovers, birds recoup what they have lost on a previous flight. Birds freshly arrived at stopover points at first gain weight relatively slowly, and they then require a high protein diet to speed up the regeneration of their digestive tract, before they can again feed quickly and lay up fat. Days before departing, while their heart and wing muscle mass is still increasing, their liver, digestive tract, and leg muscles are already becoming lighter, with mass being converted to fat. The birds act as if they optimize organ size for the risks and rewards of fueling and flight. In short, when they are being made, they require raw materials like those that their bodies are made of. Later, they are more like machines that require fuel and much smaller amounts of protein. Similarly, in training for running 100 kilometers, I presumed I'd experience wear and tear that would necessitate resynthesis of tissues. The best foods I needed for long-term maintenance to repair the wear and tear on my body from heavy training would also be protein, but protein would not be the best source of nourishment for short-term fuel in any one race. There are three reasons why I would not consider eating large amounts of protein on a run: the work required to digest it; the poisonous by-product, urea; and the loss of water that would have to be expended to flush the urea out.

Allocating specific foods for training and for racing is not an issue to runners of shorter distances, but as I will show, it was a great concern to me as an ultradistance runner. A sprinter relies almost exclusively on ATP (adenosine triphosphate) and CP (creatine phosphate), the cellular energy currency for immediate, or instant, use. He may also dip into glycogen reserves in the muscles themselves. A middle-distance runner will have to use the glycogen reserves stored in the liver, and access them via the blood to replenish the ATP and CP. Glycogen is an energy-storage molecule the body makes from almost any food containing protein, fat, or carbohydrates, and we always keep a store of it on hand. As far as short- and middle-distance running fuel is concerned, it therefore makes little difference what is *eaten,* because all food intake gets converted to the same currency of ATP and glycogen. The current 1,500-meter world record holder, who I predict will win the next Olympic gold in that event, is Moroccan Hicham Guerrouj,* who runs on couscous, a carbohydrate-rich food. Finland's Lasse Viren, four-time Olympic gold medalist, winning both 5,000 and 10,000 meters in 1972 and 1976, attributed his spectacular racing success to a diet of reindeer milk (although I'll wager his success is more likely due to generous doses of *sisu,* a Finnish entity that translates roughly to guts and stubbornness).

Problems of racing fuel arise in long-distance running because ATP stores are used up in seconds, and the amount of glycogen the body can store is about 2,000 kilocalories, whereas to run 100 kilometers requires a total expenditure of near 6,000 kilocalories. In most of us, maximum glycogen stores are exhausted even before running to near the marathon distance of 26 miles. Glycogen depletion is described as "hitting the wall" because that is what it feels like when blood glucose (derived from the liver glycogen stores) suddenly plummets. A metabolic switch-over occurs. It is not sudden or absolute, but it occurs. We start to use up body fat and protein as running fuels, or we may refuel by eating on the run. Both of these options greatly compromise speed.

I didn't know what food would be best for me on an ultramarathon, but I let data from animals point to possibilities. In bees, for example, flight endurance (given no temperature limitation) is almost strictly a function of how much carbohydrate fuel they have in the stomach, and usually that is sufficient to get them where they need to go. A bumblebee with a honey crop full to equal body weight of

*He came in second to the Kenyan Noah Nzeni, who ran 3:22.07, a new Olympic record.

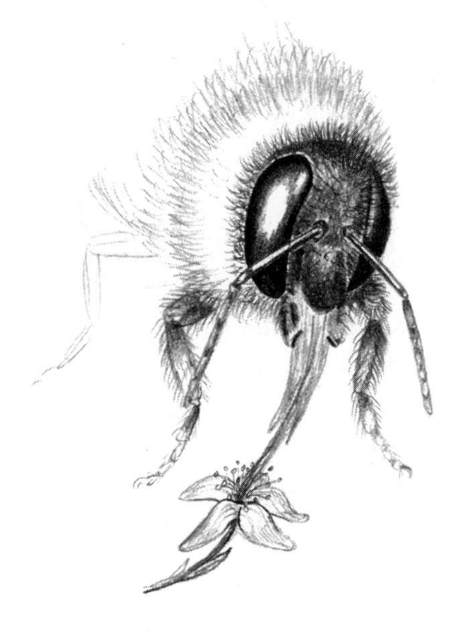

Bumblebee

concentrated nectar (30 percent sugar, 70 percent water) should have a maximum flying time of almost 3 hours. Migrating birds mobilize body fat as fuel. If a bird doubles body weight by laying on fat (but without water), it can fly nonstop for three days and nights. Thus in terms of fuel economy for long-distance travel, fat is much better aviation fuel than a watery sugar solution.

Carbohydrate is better for speed. As long as a bee's honey crop is not empty, its blood sugar remains high and normal flight speed is maintained. As soon as its crop is empty, flight stops. Similarly, when our blood glucose levels dip, we become weak. The difference is that although we may be too weak to run, we can still walk. That's because we then dip into the fat and protein reserves that our bodies save for emergencies. Although these fuels contain much energy, it is difficult for us to access them rapidly, as birds can and do.

Bee with full honey crop

Individual bumblebees and honeybees can afford to run almost totally on carbohydrate because they don't usually run out of fuel. They always have access to honey stores in the hive of which they are a part. They revisit the hive usually at intervals of half an hour or less. As a consequence, they have not evolved the need to store body fat when they are confronted with a surplus of food—they store extra calories in the hive's larder instead.

The whole idea of my trying to increase the fat mobilization into the blood and the fat-burning metabolism of the cells was to delay exhaustion by conserving the precious carbohydrate fuel. Blood sugar (or liver glycogen, and also protein, from which sugar is derived) is absolutely essential for brain function, and one solution to keep glucose levels up on ultramarathon runs is to get carbohydrate from the digestive tract. For most runners, however, the stomach doesn't cooperate well with running. It seems as if food processing and running are mutually exclusive, and many runners who try to eat and run end up throwing up. That may be an adaptive response for running fast over a short distance. Stomach contents add weight and rob precious blood supply from the muscles, making one feel and be weak. Normally the predator eats *after* the chase, not before it, and the stomach's response makes sense. But what if the stomach is trained? I trained by eating a sandwich, a hamburger and potatoes, or even a full meal immediately before a long run. It seldom caused problems, but then I didn't run fast, either.

Naoko Takahashi, a famous women's marathon runner from Japan,* claims that the secret of her success is drinking the stomach secretions of the larval grubs of the giant killer hornets, *Mandarina japonica,* both during training and the race itself. That juice originates from the chewed-up, digested insect prey (mostly bees) the adults catch and feed their grubs in the communal nests. In return for being fed, the 4,000 or so grubs per nest regurgitate clear liquid droplets to these adults, who then fly up to 60 miles per day at up to 20 miles per hour in their hunting excursions. It was this clear liquid regurgitate that Tokyo investigators tested on mice and students, claiming it boosted their ability for fat metabolism while reducing muscle fatigue and slowing lactic acid build-up. (The question is, as opposed to what?) My Maine buddies from the Rowdies Running Club would undoubtedly claim that defizzed Coke or beer is handier, and maybe better. I suspect that the wasp juice contains a very high sugar content and many amino acids (but not more than in honey and meat). As to the wasps' speed and endurance, bees do even better, and they burn more honey.

I have not yet experimented with wasp regurgitate, but I did try honey, a bee regurgitate. My experiment with honey on the run was when I lived in Walnut Creek, California, and was

*Winner of the 2000 Sydney Olympic marathon.

training for the San Francisco marathon. My long training run took me up into the foothills of Mount Tamalpais and back, in usually scorching weather. I love honey, but nearly a quart all at once was a push. Nevertheless, I forced it down and headed out the door, running through the suburban sprawl and off toward the mountain. I soon had mixed sensations, but those in the gut predominated. Almost irresistible urges. It was only with great effort that I made the cover of some bushes near the foot of the mountain. Well, I'd made it halfway, but I felt decidedly woozy. I then lost a good deal of liquid along with all the honey, and I think I subjected my body to dehydration among a few other sins. But I learned my lesson well. Until I tried an equal volume of olive oil.

My third experiment was with a combination of lots of carbohydrate and lots of water—beer. I had done my trial runs on a 20-mile course, making a beer cache 10 miles out under some bushes. I timed myself out to the beer, downed the twelve-ounce bottle, then ran on and timed myself with the stopwatch over the second part of the course. If I slowed down, I figured I'd better try something else. If I speeded up, I could be onto something. I had speeded up slightly.

For a real test, I entered a long road race toting three six-packs. Presuming a fast racing pace, I planned on having one every 4 miles. We took off like a rhinoceros in rut, and I was soon in the lead, chugging one beer after the other and increasing my lead even further. While starting to congratulate myself on the great run, with just three beers left to go, I suddenly felt weak. With two left to go, I lost all my will and just dropped out. I felt sick. More fine-tuning would not have been a bad thing if I'd really planned on this as something serious. However, I did not repeat the beer experiments. Instead I tried Ocean Spray cranberry juice. At first, I drank it only after my training runs to counter my usual loss of 4 to 6 pounds of fluids. I never carried groceries with me, because I didn't like the burden affecting my stride, which I needed to cultivate. I cached the juice along my training route so I wouldn't have to wait till the finish.

Cranberry juice is, like nectar or dilute honey, mainly a sugar solution. Why not use something more concentrated, like a candy bar or a sandwich? I tried those, too. They were great, unless I was running fast or had already run far and was getting dehydrated, which was precisely when I most needed calories. My mouth would feel as if it were stuffed with cotton, and I couldn't swallow. I can down six saltine crackers in under a minute (having done so on a bet), but during race conditions, the only way I could ingest sufficient carbohydrate was in liquid form.

Cranberry juice became both my racing fuel and my water. I decided I might as well ask Ocean Spray to sponsor my run by giving me free juice for the race in Chicago. To my delight, they not only agreed, they also picked up the tab for my airfare and hotel reservations. That's the full extent of *material* compensation I've ever received as an athlete, and it's all I ever asked for.

I know of nobody else who has tried cranberry juice since. For current advice on diet for an ultramarathon, I quote Jan Vandendriesche of Belgium, on winning Boston's 100-kilometer challenge on October 10, 1999: "Go light, but stay ahead of the calories. Don't stuff yourself till you barf. Avoid ibuprofen if possible. Carbofuel for the first half. Start taking GU (a commercial concoction) at about 40 km. Drink Metobol (also a commercial concoction) midway, then nothing except GU, water, and Pepsi (ditto) the last twenty miles."

A quick perusal of some of the products advertised in *Ultrarunning* magazine yields Hammer Gel and E-Caps Sustained Energy. Both contain complex carbohydrates plus four "key amino acids," or protein, and both are advertised as containing "NO sugar." Then there are Pre-Race CAPS that have electrolytes found in blood plasma, and Recovery CAPS with "special nutrients and antioxidants" for after the race. All of this could be a far cry from eating yeast rolls prior to the race, drinking an almost pure corn-syrup sugar solution with a strong diuretic in it, and looking forward to gorging on steak, baked potato, and apple pie topped with generous servings of ice cream immediately afterward.

EIGHTEEN

Training for the Race

Running is the greatest metaphor for life,
because you get out of it what you put into it.

—OPRAH WINFREY

It is natural for us to run, given the appropriate environmental stimuli, because during our evolution we acquired the genes that allow us to do so. I therefore have some faith that, given the right stimuli, most of us can become runners. We have the lungs, limbs, hearts, and minds for it, just as sandpipers have what it takes to migrate from North to South America. Of course, natural is not the same as good. Running well is for us now a value, not a necessity.

As Tanzania's great marathon runner Juma Ikangaa has said, "The will to win means nothing without the will to prepare." But how do you train yourself to do something so totally new as to run a distance that is twice what you've ever run before, when the distance you've run before felt close to your maximum? One thing I was sure of was that to *train,* as opposed to just run, I needed to have ideas of what I'm going to achieve and how I'll do it. Any mistakes would be, as my experiments with honey, beer, and olive oil had shown, great lessons.

One of the beautiful things about running is that it is direct and elegant. The formula is simple: put one foot in front of the other. It doesn't take much to figure out that if you want to improve sprint speed, you run faster. If you want to improve distance-running performance, you run farther. Although that's all fine and good for getting in shape, it's not good enough when you're going for a win or a record. It is a given that the cheetah is already in superb physical shape, but the outcome of its race against an antelope, which is also in superb condition, can go either way, and it is decided by milliseconds of that indefinable something extra, or a blunder. The question is, how do you achieve that extra without any blunders?

As in Einstein's formula $E=mc^2$, I'd have to juggle a complex interrelationship involving energy, mass, and velocity. My sprint and cruising speeds were presumably fast enough to be a constant in the equation; I probably could not or need not alter them. The main alterable variable and limiting factor, I presumed, would be endurance, controlled by body temperature, fluid management, and ultimately by fuel supply in the muscles, blood, liver, digestive tract, and body fat deposits and by fuel mobilization. Clearly, there were two means to extend my fuel supplies. One was to supplement with carbohydrate ingestion during the run; the other was to train the fat-mobilization and fat-burning enzymes in order to tap into body fat reserves. In humans, available carbohydrate stores are thought to be sufficient to sustain maximum work output for 20 to 30 minutes, although fat stores can sustain a high work output for days. The problem for me would be that the rate of energy expenditure (and hence running velocity) generated by burning fat is only about 60 percent that available through carbohydrate. It is thought that high and continuous rates of aerobic work are best supported by mixed and simultaneous use of carbohydrate and fat.

To run to 62.2 miles would be to enter the body's carbohydrate-fuel-depletion zone twice over. In humans, glycogen depletion spells exhaustion. To extend the tolerance zone, I trained to run on empty, forcing my body to get used to dipping into body fat as a running fuel after the muscle and liver glycogen reserves went low. I also wanted to use just as much carbo as my digestive tract would possibly process, to save those glycogen reserves for as long as possible before having to resort to fat metabolism, because carbo (sugar and glycogen) is high-octane fuel that provides greater power, and hence speed, than fat. As mentioned, sometimes I also practiced the opposite, running on a full stomach in order to be able to draw on that fuel. Runners competing at shorter distances, during which carbohydrate reserves suffice, need to train the digestive tract to shut off, as the body normally does with little inducement. Instead, I trained my gut to keep going, even as I ran.

Sometimes, to catch your antelope, you have to make compromises. The key to great ultramarathon performance is in setting goals and finding just the right balance between opposite and equally important necessities. Training must include intensely high mileage. Yet rest and recuperation are equally important. It takes rigid discipline to put in 10, 20, maybe 30 miles per day, with no excuses allowed, yet you need to be able to let up instantly when further effort might mean injury. Sometimes it's necessary to pay immediate attention to the first hint of a blister or a slight muscle tear, while at other times you've got to be able to ignore pain for hours on end. Racing mentality requires a steady, unflappable calmness, and also a devil-may-care abandon where all the stops are pulled. Success requires uncompromising logic, and subservience to an overall goal that has, as life itself, no logical basis whatsoever. As Great Britain's ultramarathon great Don Ritchie has said, "To run an ultramarathon, you need good training background, and a suitable mental attitude, i.e., you must be a little crazy."

We live in a biological world of conflicting truths that together create the ever changing new out of the ageless. Our world is not a linear-logical construct that yields truths through ad infinitum extrapolation (except in physics?) by the use of scientific tools such as mathematics. That world, the one of unbending physical truths, exists in theory, but theory tends to be just an academic exercise when it has to compete with the reality of existence, or the real world of biology that we inhabit—the one that is both incredibly finely structured and chaotic. There is no precise formula that specifies how to prepare. There are only approximations, and best approximations, until something better comes along. Like infinity, the ultimate truth can be approached but never attained.

Should training involve running with a full, or empty stomach; going on short, fast, and frequent runs, or long, slow, and infrequent jogs—or some combination of both? What would best prepare me to achieve long and fast? At that time, in 1981, I knew of no ultramarathon training manuals. I did not know how others did it, and even if I had, how could I know whether their way was best for me? I was at my tar-paper shack in the Maine woods, training for the big race by running, felling spruce and fir trees with my ax, building a log cabin, and spending time in the field with bumblebees and an owl. I wanted my training to be as pure and elemental as my racing would be. No heart monitors strapped to my chest; I could gauge effort with my own body. No stretching and weight-lifting required. No fancy shoes with baubles and bubbles, nor stretch pants and synthetic warm-ups. No pills of any sort—not even an aspirin. My only concession to performance-enhancing chemistry was a good cup of morning coffee with plenty of sugar and condensed milk out of a can. No locked-in training plan, except to run consistently at a pace and distance that I felt I could handle reasonably. No rigid protocol. Only guidelines. Too much protocol can be suffocating and become an end in itself. Each day is special. Each day is different. I ended up running mostly at near-racing pace, usually speeding up at the end to hurry home to the hut and get the workout done, which served well to deplete my glycogen stores and allow the fat utilization to kick in.

The next most important consideration in running long distance is running *efficiency*—converting as much as possible of the expended energy into mileage. For a well-coordinated human to run a mile takes about 1,600 steps, at a cost of about 100 kilocalories for a man weighing 150 pounds. As much as possible, energy should be used to gain forward momentum while minimizing up-down movement. One's feet, of course, necessarily involve an investment of up-and-down motion. But for a distance runner, that motion must be minimized in order to conserve energy. In a sprinter, energy economy per distance traveled is almost irrelevant, and knee lift and foot lift must be greatly enhanced in order to lengthen stride.

Lifting the feet a little with each stride, of course, cannot be avoided. It is just one of those costs that every runner has to pay. But animals have evolved to minimize cost. Birds minimize the lift work in their wing-limbs by having most of the muscles that move their wings on the chest, rather than on the limb itself. That way the limb stays light. Further energy is saved by having a very short humerus, and effectively achieving a very long but light forearm by having pinions that are as light as feathers, because that's what they are. All the weight is concentrated as close as possible to the body. Swifts, the fastest endurance flyers in the bird world, have particularly short humerus bones and long, thin tapering wings. They are a particularly good example of the extreme of this principle. Similarly, such cursorial (running) animals as antelopes and ostriches minimize the weight of their lower legs, feet, and digits by reductions in number and in size. Toe amputation is not an acceptable option to most runners, but since I would have to take some 124,000 steps during the course of the race, I was quite mindful of the weight of my footwear. Lifting 1 ounce of extra foot weight 1 inch higher than necessary would, for my distance, be equivalent to lifting a 900-pound weight a distance of 1 foot.

The primary way to increase running efficiency would be to minimize leg lift while maximizing stride length, and of course, using the lightest shoes possible. I practiced running using gravity and momentum as much as possible to swing my feet. A sprinter expends exorbitant amounts of energy with each step, which is essentially a leap. I needed to train a stride that would be a compromise between an energy-efficient short step, where the feet are barely lifted, and a long stride, which necessitates more knee lift. By running as much as possible at race pace during training, I hoped to cultivate that specific optimum stride for the distance I intended to run.

What a long-distance runner can least afford to do is lift his whole body up and down on successive steps. He must glide. An ostrich or any other elite marathoner exhibits almost no up-and-down motion of the head or the shoulders. Suppose a 150-pound runner goes up-down only 3 inches with each step; then over the course of a 100-kilometer run he will have lifted his 150-pound body mass a distance of about 2 miles. That's a lot of work, and it must be strenuously avoided in favor of horizontal motion.

Similarly, the energy of breathing must be minimized. In us bipedals there is supposedly a decoupling of the breathing cycle from the gait, and hence the energy expenditure from locomotion cannot be employed to also serve in ventilating the lungs. Contrary to this theory, at distance-cruising speed I coupled my arm as well as leg swings with my breaths. The number of arm and leg swings per breathing cycle and the depths of the breaths all depend on effort. But synchronization is maintained. The energy saved by such synchronization is undoubtedly very small, but it could add up in the long run. The goal is that eventually all of my body motions would be graceful, flawless, and executed without conscious will. Yet occasional editing of my form during training seemed helpful, especially when I was tiring on long-distance training runs, when coordination and running efficiency started to decline precipitously.

Training intensely would be important, but anything extrapolated to its logical conclusion leads to chaos, while anything not pursued with sufficient vigor is a waste of precious time. The Australian miler Herb Elliott claimed that his training routine under his coach, Percy Cerutti, would be considered "frightening" these days. That training was surely mostly speed work, for training the carbo metabolism needed for his specialty. I had once trained for speed myself, just barely managing to go under 2 minutes in the half mile. For me, it was a breakthrough, and it had been accomplished only after age thirty by repetitive speed work, mostly quarter and half miles, on the track. The key to ultrarunning success is obviously to run much slower, to *spare* carbo metabolism in order to last the distance. I'd never again have such sprint speed after training for distance. Nevertheless, I felt that training intensity was crucial, and for distance running that intensity would have to come from doing distance, not speed as such. Since I'd be running slower, the appropriately hard intensity could be achieved only by running to the fatigue point, and then to keep right on pushing, pushing, pushing some more: to run often in an exhausted state. By that time I endured it, because what had at first been a chore had become a ritual, and the ritual a habit. I did not have to think about it. I just did it.

As much as I believe that every twitch in my body has a physical cause and that what I do matters, there are times when imponderables rule. There were training days when I felt half dead. It was an effort to put one foot in front of the other. I slogged through the miles like a zombie, feeling aching tiredness through to the end. I wondered, Is it the pork chops I ate last night? Did I run too hard yesterday? Was it the peanut butter sandwich I ate just before leaving? There were other days when I felt light and buoyant, and I kept speeding up, feeling I could go on. On those days, instead of my perhaps intended 15 miles, I'd run 20 or 25. I wondered why I felt so good. It was always easy to rationalize. But the bottom line was, I simply didn't know what imponderables were or are at work. I just hoped that I could identify some relevant factors, and that race day would be one of my lucky days.

When I started training in early May, I ran two or three 15-to-17-mile runs on successive days with the expectation of running the race on October 4. I was surprised to find myself getting weaker rather than stronger. That was frightening. Then I figured it out. I noticed that along with my fatigue came hunger, sometimes intense hunger. There were several times that I got so weak and hungry that I knocked at a stranger's door, begging for a slice of dry bread. Once I was forced to stop and get a candy bar on credit at the village store only 3 miles from home, the cabin in the woods. The carbo worked wonders—I always made it home.

It became quite clear to me that in running distance, the big limitation was indeed fuel in the tank. I noticed that I gradually ran farther and farther with training. Why did I go farther before hitting the empty mark? My first hunch was that I was gradually training my body to burn some other fuel. That could only be body fat, although I could also have been packing more glycogen into the liver with each meal. My body would want to hang on to all the fat it could, but since it could not always indulge in its preferred carbohydrate fuel, it had to relent. The body would always use carbohydrate first, if it had any available. All of the carbohydrate stores in muscles and the liver are just barely sufficient in a highly trained middle-distance runner to finish a marathon. That would be just the *start* of the 100-kilometer, a frightening thought given that on some training runs where I went easy I felt exhausted already at near 20 miles. My second hunch relates to running efficiency. I was surprised that although I lost about 5 pounds of body weight fairly rapidly, I then soon stabilized my weight even though I ran about 20 miles per day. I never again got hungry during the run, and overall I was not eating much more than I normally did when I was not training. I was somehow getting incredibly more mileage out of the same number of calories. I suspect, therefore, that my running mechanics and possibly the cellular metabolism were becoming more efficient. That is, more energy was being converted to mechanical power and less ending up as heat.

They say that in racing one is really competing against oneself. For me, the race began in May, and until October 4 I would run it all by myself, against myself. In the meantime, I collected mileage. I could not run my racing distance on any one day, but by running often or long in a depleted state I simulated race conditions. Surgery on a broken medial meniscus cartilage of the left knee (due to injury sustained while chopping trees for the log cabin I was building) on May 19 initially slowed me down, and indeed at first almost crushed my dream. Then it made me more determined than ever. After a couple of weeks I increased my mileage again and by late summer I averaged 20 miles per day, generally alternating daily between longer and shorter runs. I regularly did a 30-mile run on the weekend, and except for two weeks around the aforementioned operation, I took not a single day off. I consistently ran over 120 miles and often up to 140 miles per week.

As Frank Shorter has noted, "There's a certain aspect of any athlete's mentality, which is to not waste an effort." Since saving time is important in my schedule, I had to rationalize to myself, and especially to others, why I could spend two, sometimes even three, hours per day, for up to three months, just running down roads. My excuse is less that I can think and reflect on long, quiet runs than the realization that others are capable of just chatting or reading the newspaper for nearly as long, and thinking it time well spent. As for loss of my scientific productivity, the time of preparation for the race might add up to what is needed to produce one small research publication that only a half dozen people would likely ever read, if it's experimental science. Anyone who is able to justify the second can surely justify the first, and vice versa. And I decided that I could.

In late August, when we came out of the woods and returned to Burlington to teach at the University of Vermont, I still did long runs, usually to and from work. Also, I now sometimes ran at noon, or put in a short run before or after the daily long run. I wanted to make my body think it had to run all the time. I never walked. Even if I needed to go only 50 yards to the library or to my car, I jogged. My body must not be allowed to think it could ever walk again. Running had to become the natural mode of locomotion, and it became such.

The motions of a trained athlete are automatic. He doesn't have to think about most of them, unless trying to ferret out flaws. I sometimes went to sleep at night trying to consciously visualize my stride so that I felt like I was feeling the poetry of motion, without the added pain. It made me feel good, as if I were a spectator and a critic. Even the tiniest inefficiencies of movement can make a huge difference over a long distance. I often noticed that muscle tenseness could be relaxed by conscious effort. I then focused attention on my calves, thighs, arms, trying to relax them even during training runs, so that only the most essential running muscles would be exercised. For a mile or so I would monitor and hence try to control the kick of my arm swings, to make sure no energy was wasted in side-to-side motion.

By mid-September I was getting close. This was really going to happen! On September 15, after a long, exhausting run the previous day, I did four short, easy runs of 2, 7, 2, and 2 miles. The next day I ran 2 miles, 20 miles (in 2:00), and then another 2 miles. The next day, 2, 10, and another 2 miles. And the next day 2, 40 miles (in 4:39), and then 2 more miles in the evening. I wrote in my training log that I ran the first 20-mile loop of the forty in 2:19, and that I was "about to hit the wall but then gulped three cartons of juice, and picked up again . . . finished fairly strong. Feel confident I could have gone much farther and faster, if I'd had two juices every 5 miles."

On September 19 I relaxed and did only 10 miles, and noted that "legs feel fine." I was on track. The next day I ran 2 miles, then a regular 7-mile loop, but fast, for a time of 41:13, which broke my previous training record for that loop by 16 seconds. My log says: "Felt slow at first and strong at end. Could have gone on for much farther at same pace." The following day, even though it was raining and windy, I did my 20-mile loop in 1:58:30, also a new course training record. It was not all out. I usually tried to keep a little back, so that willpower would accumulate, like a battery on charge.

I did not have the benefit of hindsight that the recent work on the physiology of migrating birds now provides. However, I did have an inkling that intense training for an ultramarathon has some similarities to a bird's intercontinental migrations. Birds migrate to their ultimate goal in stages, which are like ultramarathon workouts. Even as they exercise, their work capacity drops steadily as body tissues—including wing muscle and heart—are burned up for fuel. Curiously the resting metabolic rate of their remaining body tissues drops a whopping 45 percent. I did not know this at the time, but I still felt that training hard for too long would be counter productive, if not damaging to the body. I tried to peak for a specific race, much like birds physiologically prepare for each specific ultramarathon flight.

Three days later (no days off), I entered a 10-mile tune-up race in nearby Essex Junction, Vermont, not to race but to get a feel for my pace under race conditions. I did not record my finishing place (I think I was third), but I listed my time as 54:03 in the running log. The next day I took it easy, doing only 7 miles, and the following day I set a new record (by 5 minutes) on my 20-mile course, a 1:53:30, for a 5:40 per mile pace, and I wrote, "Felt strong at end. *Not* exhausted. Could probably have carried this pace for 15–20 more miles. Did not stop to drink." I was recovering quickly, as is necessary for running long and hard. Only six more days until Chicago, when I could let loose and it would all be over.

In just a short period of time, relative to what a lifelong endurance predator would have experienced, there was no longer any doubt that I could do what was unthinkable before. I could run farther, faster, and at longer intervals than I ever imagined, and if I could do it, so could almost any other healthy adult. We are all runners, if we have to be or seriously want to be. I knew I could handle the previously unthinkable exertion of running the entire 100 kilometers. The only relevant question was, at what pace?

The point of training is to condition the body to handle exertion. At some point exercise becomes a stress, physically, psychologically, and hormonally. Much has been written in the scientific literature about stress. It is thought not to be healthy for you and that you should avoid it. I wasn't worried. Stress is the expenditure of energy. You can't *live* without it. Some think that you are given a specific number of heartbeats per lifetime, and they would rather not use up their quota by running down a road. Although running at less than maximum speed may increase the heart rate up to 120 to 140 beats per minute from a "normal" 60 beats per minute, after a few months of training, the resting heart rate for the other twenty-three hours of the day may greatly decrease. (Mine went down to 34 per minute.) That is, at a daily cost of 4,000 heartbeats during an hour of exercise, the athlete may save nearly 36,000 heartbeats a day, for a net saving of 32,000 beats a day out of 86,000 used otherwise, assuming the stress of running is balanced with the *rest* needed to allow the benefits of stress to be realized.

I heard other arguments against running. One colleague who was worried about my longevity (or sanity?) sent me a research publication showing that running exercise induces the release of the stress hormone cortisone, which other studies had implicated in causing brain damage and other degenerative conditions. The researchers concluded that, therefore, chronic running would induce loss of neurons and cause premature aging. I wasn't particularly worried. As is usual, there are usually other effects as well. In another study, Barry L. Jacobs and colleagues from the neuroscience program at Princeton University showed that when mice ran every day on an exercise wheel, they developed more brain cells and they learned faster than sedentary controls. I believe in mice.

What most researchers failed to consider was the variable of time. Running is stressful only to a person who is unaccustomed to the exertion. But with training, running becomes normal. I had decided to run so much that normal running would no longer be a stress, only fast or long running. I could now run 20 miles every day and it felt quite normal to me. The important thing was that I had worked up to it gradually. Time and timing is everything. There is a world of difference between immediate effects and ultimate consequences. During training, it was less important to avoid running stress than to avoid other stresses, like those I had experienced as a student. If stress indeed causes loss of neurons, if not also brain damage, then I sustained a lot more brain damage while I was a student than when I was a runner.

I also believe in antelopes. They would not have missed a trick when it comes to running speed and endurance. I had never seen or heard of an antelope who was flexible and did stretching, or who lifted weights for extra strength. I had never heard of one doing much more than eating and running. However, if I had read modern exercise literature, such as the recent excellent book by McArdle, Katch, and Katch, or *The Lore of Running* by Tim Noakes, I might have done not only stretching, speed training, and weight lifting, but also warm-ups and cooldowns. I had, of course, heard rationalizations for all of these procedures, but I didn't know if I was hearing facts or folklore. I suspected that the most highly motivated and the best runners would try almost anything no matter how bizarre, if it was recommended for improvement on the basis of some rationale. If, for example, a world record holder always took some herb on the basis of some pseudoscientific hunch, then others would copy him and soon advertisers would crow about the product, attributing power to it, and thus, with potentially no evidence whatsoever, folklore would substitute for fact. I didn't know of any convincing set of data that proves, for a runner who puts in 10 or 20 miles per day, the specific increment of improvement that would result from a given number of stretches or power lifts. I'd never heard of a yogi who was a great runner. It seemed to me that muscles are like rubber bands—they produce power only after being pulled taut. If long and lax, they have to contract "for nothing" *until* they become taut. I never stretched.

As for running causing premature aging, I also know of no objective studies that go beyond rationalizations to provide proof. As to rationalizations, my own was that since many animal studies, from mice to monkeys, show that reduction of caloric (food) intake increases health and longevity, then maybe the *relevant* variable is caloric *deficit,* and so exercise, by reducing caloric excess, should do the same as food reduction.

It is very difficult to factor out the many lifestyle variables that are correlated with people who run as opposed to those who don't. For sheer longevity, it is hard to beat Carlton Mendell, a competitive runner since high school and a former member of the Maine Senate and the Maine Rowdies running club. At age sixty-two Carlton covered 125.5 miles on the Bowdoin College track to set the U.S. age-group record. By 1999, at age seventy-eight, he had run 126 marathons, and had no plans yet to shy away from running—not even ultramarathons. I'm not sure if his continuing vigor is due to his running or in spite of it. I suspect it is a little of both. He is an experiment of one. And so, with respect to training, am I. The results of that experiment were about to be revealed.

NINETEEN

Final Preparations

God has given me the ability. The rest is up to me. Believe. Believe. Believe.

—BILLY MILLS, *an Oglala Sioux from South Dakota, in a training diary entry prior to his surprise upset 10-km victory in the Tokyo Olympics, when he also set a personal best for the distance by 46 seconds.*

To finish the 100-kilometer in 6.5 hours, near what I hoped to do, would mean holding an average pace of 6:17 minutes per mile for the whole 62 miles. The magnitude of that task seemed just shy of intimidating. On the other hand, I thought I could do it, and I did not want the 1,350 training miles, starting from 10 miles per week after the knee operation, to over 130 miles per week during the last five weeks, to be a wasted effort. I would not be racing again next weekend, possibly ever. Only quality mattered, and to do it once and do it right, little could now be left to chance. There was no room for error. To intend not to make any error was not enough. It was a matter of will, to make sure that every damn thing that *could* go wrong didn't, and that everything that could go right did.

I had already gambled on the training regimen, the intended carbo/liquid intake during the race, and now I made one more gamble: I decided to submit to the stressful carbo-depletion and carbo-loading regime. The theory behind it is that if the liver and muscles are totally depleted of glycogen, then they will overcompensate and take up more than they normally hold when given an abundance of dietary carbohydrate. It is like gorging after a fast, nature's way of making up for previous deprivation.

The carbo-depletion regimen is risky, because it makes you vulnerable to the flu and other infections. Although I could have an excellent run even without the tiny potential boost of a few more calories of glycogen in my muscles and liver at the start of the race, that tiny amount also might make all the difference. I had to believe.

There was another reason I risked going through the carbo-depletion, carbo-loading regime. I decided to deplete my carbohydrates not only by exercise, but also by switching over to a strict diet of fats with a little protein. I hoped thereby to give an additional boost to my fat-burning machinery.

Burning the fats from adipose or skin deposits depends on adipose triglyceride lipase, an enzyme that is hormone activated. To activate it from the inactive, 'b' to the active, 'a' form requires adrenaline, glucagon, and noradrenaline, together with a drop of insulin in the blood. Training, starvation, and diet can all double or triple lipase activity, and in addition I could expect the excitement of the race to be highly relevant for my hormonal profile as well. The increased lipase activity from adrenaline release would result in reductions of glucose utilization, to extend the glycogen reserves and to postpone fatigue.

Not all ultrarunners believe in carbo depletion. When asked what diet he used on a 50-mile race, ultrarunner Jim Pearson said, "I once got beat by a guy eating a peanut butter sandwich, and he wore a T-shirt that said on the back, 'I'd rather eat worms than deplete.' " After I'd eaten little but butter, peanut butter, and cheese and fatty meat for a couple of days, I felt nauseated, half wondering if maybe eating worms might not be preferable after all. (Nor would I recommend what I did, in retrospect. The next two times I tried it I got a flu and had to spend days in bed just before the race and consequently ran very poorly.)

According to my running log I "ate well" on the night of September 28, after that day having set a training record of 1:53 over my 20-mile course. I felt strong to the end and "could probably have carried the pace for 15 to 20 more miles." Since my pace averaged 5:39 per mile, I felt that with race adrenaline, I could potentially be within range of my intended 6:17 per mile race pace for the full 100 kilometers, a distance more than twice I'd ever run before. But there was no way of knowing if I could hang on after entering the unknown territory past about 30 miles. I had put on the pressure in running this particular 20-mile run to speed up the glycogen depletion process, which I hoped to accomplish fully the next day, September 29, six days before the race.

The final depletion run on September 29 was indeed scary. I ate nothing but cheese and peanut butter for breakfast, and then set out to run my 20-mile training loop twice without even stopping, except to drink water at the halfway point. I started at a leisurely 7:00 per mile pace. At 25 miles I was "getting weak" and at 34 miles I still thought I could make it "with guts." But at 35 miles I wobbled badly and then had to walk. This was it. I'd bonked.

It doesn't take too many bonkings before vapors of doubt threaten to suffuse the brain and anesthetize the will. After all, my goal was not just to survive the distance, nor to jog it. It was to *run* the whole way, and the only thing that really mattered was speed. However, I'd been here before. It had happened again and again, week after week, that I would run a small fraction of the total race distance in training and reach a state of exhaustion where I came to a crawl. Those times were difficult, because the physical feeling of exhaustion came with the mental realization that, at that point, running 5–10 more miles would be quite impossible, and *race* distance might be well over 30 miles longer. At times like that there came the temptation to give up, to tell yourself it's impossible. But somehow you keep going, mostly thinking you'll do all you *can* do, and leave the rest to fate, hoping for improvement by race time. This time, though, there was a slight mental compensation to the exhaustion. I'd *wanted* to deplete, and now I knew that I really had *achieved* that state. In retrospect I would not do it again this way. The current lore, based on much scientific study, is most recently explained by Ed Eyestone in the August 2000 issue of *Runner's World*. Rather than increased mileage with less than a week to go as I had done, there should be a tapering, or gradual reduction, of mileage of about 20 percent per week during the three weeks before the race, with "the length of the taper depending on the length of the race you're preparing to win." I'd obviously not read the instructions.

I maintained the fat diet for only one more day, and on the morning of October 1, after a 9-mile run, by now presumably fueled only on fat, I felt "pretty good." That was encouraging. I was getting over my fat nausea. I then allowed myself to gobble bread, spaghetti, cookies, cereal, starting at breakfast, to carbo-load liver and muscles.

There were now only two more days with nothing to do but eat till the race. But should I stop running entirely? Up to this day I'd told my body it must always run. I did not want to shock it with inactivity, or permit it to forget running. I had no way of knowing, but my hunch was that an ultraendurance predator that runs to depletion in order to catch a huge feast would have *cycles* of energy expenditure with sharp switch-overs of energy conservation once the chase ends. Since for me the chase had not yet ended, I still jogged lightly twice per day, even on the evening before the race, holding myself to a very slow pace and doing only 2 to 3 miles at a time; I hoped to conserve my preciously accumulated carbohydrate fuel. Finally, I felt intense relief—there was nothing more to do. I thought I had done everything possible.

Curiously, throughout the last three months I had an image of myself as being too hefty. However, my log for October 1 says that before the 9-mile run, my weight was 142 pounds. (It went up at least 5 pounds after the carbo loading.) This was the lowest my weight had been in all of my running career, including high school. It had also been my only carbo depletion.

I mention my low weight because it contrasts profoundly with the mental image I had of myself. A year later I saw photographs taken at the time and I could not believe they were of me. I "knew" I was not that lean. I was absolutely convinced that the photographer had used a trick lens. I also had a negative opinion of my running ability, always feeling I run much too slowly and not far enough.

Margaret, my wife at that time, and I flew to Chicago the day before the race and we settled into our hotel near the starting point of the race by the Lake Michigan shoreline. I went out for a short jog to find the starting line in the evening so I could find it easily the next dawn.

The race organizers had scheduled a prerace gathering. Presuming there would be speeches by dignitaries and name athletes, I did not attend. It would have cost me adrenaline, and I'd need to conserve it all for the race the next morning. I had no idea who was there, but I learned later that at the prerace seminar Don Paul from San Francisco had predicted "an exciting race"; Klecker had returned, presumably for a shot at the 100-kilometer record. There was also to be a carbo-loading dinner, but it was scheduled too late. The race would start at 7 A.M. and I wanted the bulk of my big carbo meals (lots of spaghetti I'd brought along, already cooked, in a jar, and yeast rolls) already voided by 6 A.M. I needed to avoid the necessity of running for 62 miles with a full colon, and any stop along the way might undo everything; I had timed my bowels and I knew that I needed to have supper *finished* before 5 P.M.

If one wake-up system failed, I had a backup or else I would not have slept. The arranged hotel wake-up call came shortly after my alarm clock rang. I got up early to have a couple of hours to digest my breakfast of yeast rolls, and to drink the coffee we'd brought in a thermos. After breakfast, we checked out of the hotel and brought our travel bag with us to the starting line, ready to fly home right afterward; I needed to get back quickly, to get back to my work.

I was wearing my Nike Mariah running shoes. I had chosen these narrow shoes because of their light weight. To reduce weight further, I did not wear socks. Socks absorb sweat, adding more weight. I had cut small ventilation holes into the shoes with a razor blade. It does not take much to turn one's feet to hamburger when they pound the pavement without interruption for six to seven hours. Without socks, even the smallest stitch that protrudes on the inside of a shoe will do it to you. I had toughened my feet by running many training miles in these same shoes while not wearing socks, so I hoped to suffer minimal damage. My shorts had also been road tested, but just to be sure, I rubbed Vaseline onto the inside of my thighs. Sweat with constant friction can result in distracting rawness, possibly involuntarily affecting stride.

It was still dark when we got to the starting line. It was almost deserted, but then I saw several other runners jogging in the shadows on the dark raised sidewalk, against which the lake splashed in the dawn breeze. Had I really come to the right place? Soon enough more people started to arrive. Like me, they all had their private dreams. They had prepared long and hard, and they had come from all over the United States and Canada to put their dreams to the test—be it to finish the distance, to run a personal best, to win, to show, or to set a record.

Jack Canney, my handler, met me near the start with a crate of cranberry juice. I gave him my plastic squeeze bottle and instructed him to fill it and have it ready at designated spots along the 10-mile back-and-forth loop, so I could get a drink at least every 5 miles. I'd drink as fast as I could on the run and drop the bottle; he'd then retrieve it for the next tank-up. We practiced a few handoffs running at anticipated race pace.

By 6:50 A.M. there was a huge and growing crowd that would swell to 261 starters. As reckoning time approached, a few runners still did warm-up strides on the sidewalk in front and in back of the thin white line drawn across the pavement. Then as they clumped behind the line, they stretched or jogged up and down in place. The sun had come over the horizon. There was a brisk wind. We shivered slightly in the cool morning breeze.

I debated whether to keep on my tattered dark blue cotton sweatpants and sweatshirt, and decided in the last seconds to strip off the pants despite the cold, because it would take seconds off my time to do it later. The shirt could be flung off at any time without losing stride.

We packed up ever-closer behind the white line. The antelopes, world record holder Klecker, Paul, and the other fast ultrarunners, lined up directly in back of that line. The tips of their running shoes practically touched it. I, a complete unknown in this crowd, hung back a little behind them, but I squeezed in well toward the front. I became strangely calm now, perhaps relieved. I had finally arrived at the start. It would all be over in just a few hours.

This experiment of one will be, in the parlance of science, an anecdote. Nevertheless, it is still an experiment, not just a random happening. It is an experiment because I've been guided by logic derived from a vast body of experimental work on animals, and backed up by my own experiences. I've tried to incorporate the empirical facts and experiments toward achieving a specific outcome, and I have the satisfaction of knowing that I've done all I thought I could do.

With seconds to go the race director gave a speech about the course, the rules, the aid stations. . . . Some of the runners were quiet; others still anxiously jumped up and down, stretched, or slapped their arms back and forth across their chests in nervous anticipation. I reached my hand out to a man next to me: "Good luck!" I wished him. "You too!" he answered. Then I bent down once more to check my shoelaces, to make sure that they were tight and double-knotted, yet loose enough not to constrict the feet. It was so important to get them just right.

TWENTY

The Race

You don't run against a bloody stopwatch, do you hear? A runner runs against himself, against the best that's in him. . . . Against all the rottenness in the world. Against God, if you're good enough.

—BILL PERSONS, *fictional coach in Hugh Atkinson's* The Games

"On your marks—get set—" The sharp report of a starter's gun sounded. I'd heard it hundreds of times in over two decades past. But this time was different. I was forty-one years old. There would not be another chance. That's what made this fearsome. My one comfort was that I'd promised the beast in me that this would be my last effort, and my best.

Klecker and the others up front took off like antelopes pursued by wolves. Keep calm, keep calm, I kept telling myself. Don't get sucked out. I've got to run my race, on my schedule. Trying to hold myself back, yet going fast enough to hit about a 6:15 per mile pace, took concentration. I had to be like a camel, although a very fast one. I had to make sure I started slower than the pace of most of my training runs even while factoring out possible adrenaline effects that would make me feel light at the start. I would check my pace by listening to timers along the course who'd holler out the elapsed race time at the mile, 5 miles, 10 miles . . .

We had barely taken off, it seemed, when I heard the first mile split—several records over 6 minutes—just a tiny tad too fast, but right about where I wanted to be. Ray Krolewicz, the indestructible ultramarathon camel, was alongside and tried to make small talk. I could neither listen nor talk, and I soon passed him. A big crowd was ahead, and after only 3–4 miles I had lost visual contact with the leaders.

Jack came alongside me at the first-aid station to hand me my first drink. I grabbed, squirted the juice into my mouth, dropped the squeeze bottle, and kept on, trying to keep the same pace. The key to winning this would be: never speed up, never slow down, don't stop till the finish. Most important in a race, never get sucked into someone else's program. And, oh yes, *believe.*

I'd drink all I could force down. I had to rely on fat metabolism for endurance, but I'd burn carbohydrate for as long as possible—all of what I had loaded into muscles and liver and all I could process from the gut along the way—to boost my speed. Still, I worried that I might drink too much juice, because I had no way of knowing how fast I was losing liquid from sweat. It all depended on pace and temperature, both of which seemed to be rising quickly.

At the 10-mile mark I anxiously listened as the times were again called out as we rushed by. One-o-three ten, one-o-three fifteen . . ." At sixty-three minutes and 10–15 seconds (at 6:20 per mile) my time was almost a minute slower for 10 miles than what I was aiming for. A little alarm bell sounded in my brain—but the rational centers said, better slow now than later. Like antelopes, Paul and Klecker had already passed through the 10-mile mark a full 8 minutes ahead of me, and I had not been loafing! Such an incredible lead—but then, since last year Klecker had been the best ultrarunner at that distance in the world, ever. What could I expect? Some of the best runners' strategy, which the great late Pre (Steve Prefontaine) was famous for, is to try to control the race and lead from the start, intimidating the competition and then hanging on by sheer guts. Obviously not my style.

I speeded up only a tiny bit to try to get back on an even pace. I did that surprisingly successfully, because I later (eighteen years!) learned that I finished the first four 10-mile loops in 1:03:16, 1:01:31, 1:01:33, and 1:01:03.

At the end of the first 10 miles my mind was still clear, and I could concentrate on the running machine. There is always plenty of time for taking inventory of one's otherwise automatic responses on an ultramarathon. I still had 52 miles left to go.

Keep those thumbs up, I told myself, to avoid floppy wrists. Motion must be forward and back, minimizing up and down. Keep the knee lift to a minimum. I'd lock my mind on the movements, checking them out to make sure I was feeling smooth. Now—loose, loose—keep it loose. For a half mile I continue my mental tour of the running body, *visualizing* loose thighs, calves, arms . . . Then a focus on my legs. Creeping tiredness can cause inefficient motions. Relax. Swing each leg far forward while at the same time loosening all the muscles; that way only the relevant ones do the work, and none work against each other like a bumblebee's shivering response, which burns energy to produce heat but accomplishes no work. I visualize matching the stride of my right leg in synchrony with the swing of the left arm. For a quarter mile or so, I feel the rhythm and switch my conscious editing to concentrate on the opposite limbs. Once in a while, I vary the length of my stride, to contract my leg muscles for slightly different durations, like frogs varying the length of their calls. The rhythm of my footsteps is steady, unvarying, and like my heartbeat, it is unconsciously timed with my breathing. As long as I'm at rest, I can normally feel the beat when I switch on my conscious awareness. There are either one or two beats with each inspiration, and the same number with each expiration. During running I feel my breathing. Like my heartbeat, the breathing rhythm is usually also unconscious. It is timed to the same unconscious metronome that times the footsteps. I like the feeling of the strong, steady rhythm with everything in sync. At times I listen to it—in just one instant I can bring it up on my screen of consciousness. Three steps with one long inspiration, a fourth step and a quick expiration. Over and over and over again. My mantra. My mind goes blank. Sometimes I vary the beat, with only two steps during the inspiration. I can do this switch-over consciously, but it normally occurs unconsciously as I increase my effort. The harder the effort, the deeper the breaths, until there is the sudden switch in number of breaths. The rhythm preserves synchronicity, synchronicity translates to smoothness, and smoothness means energy efficiency. The body's metronome has been fine-tuned by more tens of thousands of miles than I can begin to comprehend, which have long been deleted from my working consciousness, just like the breathing rhythm itself. Only the *feeling* of it remains. And it feels good.

Meanwhile, Klecker's lead is continuing to grow ever wider. I'm almost to the marathon mark and about to grab my juice bottle when I hear Jack yell: "He is heading for a new world record." That would be in the 100-kilometer, I presume; he has already set his astounding world record in the 50-mile, last year. "You couldn't reach him now with an airmail postage stamp," Jack continues—just as I drop the juice squeeze bottle—perhaps to forestall later disappointment or to prevent me from overreaching.

I will remember his comments for as long as I live. It took some wind out of my sails. Klecker was indeed intimidating. I knew then that if I couldn't beat him, even in this one race, then I obviously couldn't have a U.S. record, either. To try to succeed, I needed to start off with fantastic dreams, but ultimately I'd have to be realistic. I tried to console myself with the thought that I must simply do the best with what I have. Giving it my all, that's all I can do. That's all I *can* care about, and therefore that's what really counts. Nevertheless, things can happen. Any weakness or flaw—no matter how slight—will be magnified in the next 10, 20, 30 miles, and after that it will be hell for all of us. That still gives plenty of room for the unseen, the unplanned, the unanticipated, in any of us. I recall Bert Hawkins, twenty-two years ago at Hinckley, Maine . . .

My time at the marathon mark was 2:42. I'm on pace again, although Klecker, of course, is far out of reach. I'll not be trying to outsprint an antelope. If I tried to be too brave, I'd overreach. I'd blow up. I'd end up a casualty alongside the path. Once you speed up to the point that you're breathing hard, you dip too deeply into the carbohydrate stores, and possibly pass the anaerobic threshold, when lactic acid is produced faster than your metabolic and cardiovascular systems can get rid of it. Lactic acid is like sand accumulating in the gears of a car that soon bring all to a grinding halt. Don't speed up, don't slow down, and above all, never stop . . .

Soon enough it gets harder. Much harder. But I'm not sure anymore whether I'm slowing down or overcompensating and possibly speeding up. I just keep working harder and harder, and steadily going by one runner after another, using them to cinch myself along. A runner up ahead—move up on him—pass—the next one.

Coming by Jack again. He yells: "Paul has dropped out—Klecker is still going strong."

The cranberry juice is becoming increasingly hard to stomach. I lose both my thirst and my appetite and have to force myself to swallow. Fatigue to the point of pain is overwhelming other sensations. My body is screaming at me to stop, and it would always win if it did not have a mind to play tricks with it, boss it around, and delude it.

To psych oneself up takes self-delusion. That's where the use of logic comes in. Logic is less an instrument for finding truth than a tool that we use to help us justify what our lower emotional centers direct or demand. Lacking this self-delusionary logic, we would be less able to rationalize, and so be unable to succumb to such mad, senseless, crazy things as trying to see how fast one can run 62.2 miles without stopping. Ultimately, our logic may get whacky enough that we see through our rationalizations, and then they don't make sense anymore. This almost invariably occurs sometime around halfway through the race, and you ask yourself, Why am I doing this? Why am I here? *Why?* There is no answer.

At that point, one needs faith—a combination of ignorance, deliberate blindness, hope, and optimism. It defies logic yet makes us able to strive and to survive. Maybe it also distinguishes the mind from a computational machine. It's what made our ancestors chase the antelope on and on till it tired.

"To run a good ultra-marathon," the world's best ultra-marathoner, Don Ritchie, has said, "you need a good training background and a suitable mental attitude—i.e., you must be a little crazy." I had the first. But the second? I ask myself: Is there anyone else in America who might be an even greater lunatic than I, who might push himself even harder? A small voice says, Probably. So I push again, a little harder. Am I crazy? Perhaps. But I must judge both my and others' ability accurately, maintain absolute integrity to any vision, and be guided strictly by cause and effect, by empirical reality. As Yogi Berra said about baseball, "It's ninety-percent mental. The other half is physical."

After about 4 hours, the sun is blazing hot and the wind has picked up. Jack still holds out the welcoming squeeze bottle full of cranberry juice every few miles. I grab, squirt, swallow, drop the bottle, and am glad to have both hands swinging free again . . . "Klecker is fading," I thought he said as I went by. What? Really? Fading? Did I hear correctly? I'm still passing people. Another 5-mile lap, another cranberry juice. "You're in second place . . ." So what? I think—he is still miles ahead.

Every handoff is now a welcome event. Every handoff ticks off another 5 miles completed.

I don't really feel thirsty, but I drink anyway, because on my training runs I had often had a lot of fluid weight loss without feeling thirst. Thirst arrives too late.

"He's dying!" Jack says at another handoff. I *did* hear it right. I think of Billy Mills as he, a complete unknown, is coming out of the last turn charging into the lead of the Olympic 10,000-meter, saying over and over to himself, I can win, I can win, I can win. I feel a shiver all over my body. Jack's two words have an electrifying effect. Although my body is weakening, I'm carried along now by what amounts to spirit. I don't really know what spirit is, but I feel different. I know I've got a chance! The impossible. Nobody knows me. And I'm charging out of nowhere, and I'm going to catch him! I know what he is experiencing. I know exactly what it feels like to die on the run. Last week was my most recent reminder. There is no way he'll go beyond the 50, and there is no way I won't.

It is time to get excited. It is time to squeeze out the adrenaline. I think of Mike, Bruce, Fred on our cross-country team, mock-growling, laughing, and saying on the run: "Gotta be tough—be an animal." I also know that being ahead means nothing—not till the finish line. Two years earlier, while I was laboring about a mile from the finish line of the San Francisco marathon, I heard snatches of chatter from radios held by spectators along the course. I recall hearing: "Here comes the winner now . . . it is Peter Demaris." Demaris was just barely visible, far ahead of me. But I wasn't ready to cede defeat, despite the premature announcement of his victory. I speeded up, and just kept going—I don't know how or from where it came, but something made me fly. My finishing time was nothing to brag about, but the *San Francisco Examiner* the next day (October 29, 1979) headlined, "Tricky ending to Golden Gate Marathon—complete unknown wins." I was that unknown runner. And I know there are many. They *can* be any of us. I might get the lead now—today—but I, too, could be caught . . . just as Demaris was caught then, right before the line.

During exercise, the mobilization of fuel for use by the muscles is controlled largely by the circulation not only of adrenaline but also of noradrenaline, adrenocorticotropic hormone, glucagon, and thyroxine—all these hormones are controlled by feedback loops through the brain. I obviously don't think of hormones or their feedback loops now, but abnormal performance demands abnormal physiology. How do I change from being normal? I involuntarily draw inspiration and strength from the example of brain power exhibited by others I admire.

Lefty had it. Lefty Gould, who had talked to me for hours when I was mail boy and when I had jogged back and forth twice a day carrying the leather mail pouch. He had looked unblinkingly at me with his pale gray eyes and told me of the "Krauts" who crawled over the lines and held him up at gunpoint, to demand his cigarettes, then let him go. In my mind's eye, I see him next in the hospital as a group of interns wheels him along. Feebly and laboriously he hoists himself up on one elbow and declares with a grin on his face, "I can lick all of you fuckers!" to show them not to pity him, the once great prizefighter. His brave but feeble gesture was a declaration of his generosity—how much he had in him, even when his body was helpless. Not to give an inch is to give everything. The stories he told me, a little kid . . . He'd lean out the teller's window to tell me how he'd thrown his shot-out thighbone at the enemy . . . as he continued shooting even as he sunk into unconsciousness. I draw deep now, on memories—on life, really. The mountain of life. Let there be enough of it to draw from . . . till the finish line . . . Lefty—now I see him—as he lay so still with his hands neatly folded across his U.S. Army uniform, in the casket in Moody Chapel, where I'd been hundreds of times as a kid before running off into the woods on Sunday afternoons. Tears well up. Lefty had been a personal bodyguard of General George S. Patton, a 1912 Olympian. Patton had said,

"Now if you are going to win the battle, you have to do one thing. You have to make the mind run the body. Never let the body tell the mind what to do. The body will always give up." The body can handle only little steps. The mind can take great leaps. To that tree . . .

My entire life is compressed into this little life of several hours. The past, present, and future fuse into a searing knot where the body shrinks and the mind is of ever-greater significance. Horizons shrink. I'm dipping into pain, gradually, inexorably. I look up and focus on a tree up ahead. Keep the pace—*to that tree*. I'd make it and reward myself with a congratulation: I *made* it! There—now to *that* tree. And so it goes, one little section of the distance at a time, distance I'd never in my life have to run again . . . ever . . . never again. *Now* is all that matters. Now—now. This *moment*.

It's as if I've been in the woods a hundred years, trailing a big whitetail buck forever—I've finally come close physically. I can do it! Don't mess up now. This is the end of the longest hunt, and the biggest buck of all is up ahead. He's that white line across the pavement. My mind locks on, drawing me forward.

Finally, as I'm nearing the end of the 50, Jack hollers excitedly: "Klecker is *finished*! You are first now, and far ahead of the next runner." Only 12 more miles to go. I grab the cranberry juice and speed on, all the more possessed. If I can finish first, that's all the more reason to run ever harder, because *now* I have a chance to set a record—if I can only hang on. But anything can happen—a muscle pull, the dreaded wall, dehydration, an ankle sprain . . . a complete unkown. . . .

The end of this hunt is drawing close. The quarry I'm trying to catch will be defined by a number, my finishing time: hours, minutes, and seconds separated by colons. And that number once made will be with me for the rest of my life. Maybe it should go on my tombstone. After all, two sets of numbers designating birth and death dates say little about a person. It is the in-between that matters. The number I am making now is pure. It will define the limits of my animal nature—it will be the measure of my imagination, achieved by gut and spirit. It can't be bought, traded, or achieved through leverage. All other honors are paltry in comparison. It is valuable, because it's a product free of others' judgments, prejudices, jealousies, and ignorance. This is life not as it is, but as we idealize it.

Don't forget the precious past. I'm forty-one. I'm nearing the end of the cat's proverbial nine lives as a runner. The back, the two knee operations, the orthopedic surgeon saying, "If you don't stop running, I'm going to have to take that kneecap off and throw it in the garbage can."

"How many miles do I have left if I stop running?" I wanted to know.

"I can't tell you that. It could go tomorrow, or it could last you twenty years."

"What if I *don't* stop running?"

"I can't tell you."

"In that case," I had told him, "I'm going to run like hell and get into the best shape of my life, and use those miles to their utmost." And I did, and won first master in the Boston Marathon. (Inactivity never helped me: this was only one of four very similar scenarios that I encountered in over forty years of running.)

A picture pops up in my mind of the tall man I saw walking across the campus at Berkeley. His face—was missing. Burned off by napalm? All of those brave soldiers—many of my running mates. Someone else went, because I didn't. . . . Their heroism—and I'm complaining that I'm *tired*?

I inhale deeply. I suck in the fresh, clean air off the lake. I run past strollers on the racecourse sidewalk who see us runners in deep concentration. We look neither right nor left. We stink of sweat. Our glazed eyes stare ahead . . . Runners don't have to smile. We don't have to look pretty. We don't allow ourselves to be judged, and we cringe every time we see a superb athlete such as an Olympic diver, gymnast, or skater having to stand to be judged, waiting for cards with numbers to be held up.

One little distance at a time. One step. Every one is precious. Every step is aliveness, because aliveness is to resist inertia. I draw on all the emotional wells I can think of, trying to kill the demons of indifference that say, Why? *Why?* It doesn't really matter—why should I care whether I win or if somebody else does? And who cares whether I finish in 6:30 or 6:31 or even 8:30, if I'm first? Nobody will know the vast differences. Except me.

I'm still passing people on successive loops who are miles behind me in the race. Bystanders can't tell who is up front from who is way back in the pack. Just as in real life.

"Suffering is the sole origin of consciousness," Dostoyevsky wrote. My stream of consciousness alternates between vividness and dreamlike somnolence. Sometimes it retrieves peaceful scenes—the antithesis of what I'm experiencing. I envision myself lying in the grass by the cabin, paddling with my pal Phil down Bog Stream, fishing with him as a teenager at dawn on Brailley Brook in the great north woods of Maine. I try to feel the birds, the forest, the people that are in me. I call upon these riches through images in my mind that "flicker about in dreams or can be called upon to relieve an hour of stress or idleness," as Howard Evans, a biologist friend, once wrote. I call upon those images now . . . I see Phil when I ran the state meet at

Bates College. He'd driven thirty miles from Wilton, the little town with the roaring, clanging textile mill where he'd gone early each morning most of his life, carrying his black lunch pail with a thermos and a sandwich. He'd driven all the way to Lewiston just to see me run. Nobody had ever done that before. I'd won it for him, against the best distance runner in the state, and in his excitement at seeing me win, he jumped out onto the track—and threw up! I'm running for you now, Phil . . . I see you in bed. Cancer has melted your frame. You can barely move. "It's a hard row to hoe," you say wearily. I have hoed hundreds of rows of beans and corn, many for you in your garden when I was a kid. "Take me to Bog Stream in the canoe," you feebly begged me. You wanted to go *out,* on our favorite canoeing stream, where we'd found peace together.

Now during the run I tap into another emotion: shame. Did you want us to tip the canoe in order to drown yourself? I had played dumb. It took you another two weeks of agony, lying staring at the ceiling, before you expired. How precious you'd think this moment is now, if you'd had it, ever. How precious I'll think it is, when I'm where you were. . . .

My heart pounds. When will it be my turn? I try next to distract myself with pleasant images—painted turtles slipping off half-submerged logs on the Kennebec River as we glide quietly around a bend with the rowboat, down past the gardens at Good Will. Bumblebees buzzing on the blue pickerelweed blooming along the banks, where the lily pads float and the pickerel lurk . . . Will my kids see these wonders? Feel them? My daughter Erica—only ten years old—has just left with her mother to live back in California. Erica—Erica—I love you—I love you . . . And my body shudders. Winning is not enough, I tell myself again and again. I had for months forgotten what the record was, after

I knew I'd try for it, because ultimately it didn't matter. My best did.

The miles roll on. My first place finish *feels* assured, but so little in life really is.

My pace now has to be maintained by a different body. The very landscape has changed. The distance between trees has expanded, the ground has hardened, the scenery is fading. There are no more bystanders. Nothing but pavement 10 feet in front of me, 5 feet, and my mind's constant vision of the quarry ahead—that white line across the pavement. The universe is contracting, constricting. The pavement—and the line—is all there is. I've run several times around the globe for this opportunity, and I could still miss it by a second. If I don't run those 100 yards to that next turn as fast as possible, I will later experience a pain greater and more long-lasting than what I feel now . . . To that tree. . . .

I remember only short phrases of songs. I had rehearsed a Cat Stevens song in hopes of distracting myself on the run: "Summers come and gone / Drifting under the dream clouds / Past the broken sun / I've been running a long time, on this traveling ground." I needed stronger and stronger medicine. Images of . . . the cabin in the woods, the tranquillity of the trees and the songs of the birds at dawn, the thoughts of their epic migrations, the feel of wet dew on the grass early in the morning, the humming of the insects on the rhododendron in the bog—the flight of the ducks last spring and their excited quacking in the swamp . . . Memories, distractions, remembrances, and the longing for "the peace beyond understanding" rush me onward.

As I'm rounding the bend with only 2 or 3 more miles to go, I'm elated by one thing, there is one overpowering, delicious thought: This *will* end soon. I speed up slightly, catching a second wind for the anticipation of release—in minutes—seconds. That foretaste of relief drives me as hard as any other motives. However, even when I finally see the finish up ahead—that group of people—I continue to fear that possibly another runner might be creeping up behind, having saved it to the end, to rush by to take me by surprise. Or if I'm close to a record, any number of them could be *invisibly* beside me, separated only by a date.

Soon I see it, up ahead—the prize, the white line across the pavement. A hundred feet—50—10 . . . Finally . . . It's over . . . I've done it! I've come through—into a heaven where merely being there is the sweetest ambrosia.

One would think I'd have raised my hands in triumph and pranced about like a mad banshee. However, I was much too exhausted to raise even a finger; instead, feeling a deep, quiet, warm glow, I collapsed onto the soft, cool grass in the shade of a tree. I felt unimaginable contentment as my heart pounded a long time from the hard finishing sprint.

The tall, solid metal winner's cup is inscribed at the base with the words "Winner—100k" and beneath them, "1981 National Champion." The cup is full. It contains what I had put into it. Like catching an antelope, the best things in life that we can experience are served on the challenge to endure and to overcome in the long run.

Bernd Heinrich crossing the finish line,
October 4, 1981, Chicago, Illinois

Epilogue

Things in motion sooner catch the eye than what not stirs.
—SHAKESPEARE, Troilus and Cressida

Every long-distance race, like every hunt, is different yet also similar in many ways. In most ways this one was like most others. After finishing the race I felt like a birder might who has ticked off another rare species for his life list.

Like trying to sight rare secretive birds, ultrarunning is not for everyone. Although *Ultrarunning* magazine voted me the outstanding male performer for 1981 on the basis of the Chicago race, the *Chicago Tribune* paid scant notice to it. While researching for this book I read their account of it on Monday, October 5. The paper focused briefly on Barney Klecker from Minnesota, and Sue Ellen Trapp from Florida, as the respective male and female winners of the 50-mile. My name was not mentioned, although I finished the 100-kilometer 4 minutes and 37 seconds before Trapp crossed the same chalk line in her 50-mile win. The oversight is unimportant, but telling.

Ultrarunning is, from a media and a spectator perspective, as one astute observer remarked, "about as exciting as watching paint dry." Actually, that doesn't quite capture the reality. It is as physiologically and psychologically challenging as any sport on the planet. But it can't be reduced to sound bites of seconds or minutes. That's why I felt challenged to write about it. I chose this race because it was my most memorable one, and most inspired. I remembered thinking while running it that in twenty years I'd write a book about it. It seemed an odd thought then. But when the twenty years were almost up, I remembered, and picked up my pencil.

I had finished the 100-kilometer race not in a dead heat as I had long imagined to myself. Instead, perhaps boringly, I crossed the finish line nearly three-quarters of an hour before the second 100-kilometer finisher. My friend Ray Krolewicz was third. My official time for the 100-kilometer on this nationally certified course had been 6:38:21, beating Frank Bozanich's North American 100-kilometer road record by 13 minutes. In the nineteen years since then, my record has been bettered by four North Americans. Most recently it was lowered to 6:30:11 by Tom Johnson. I had gone through the 50-mile point in second place behind Klecker, in 5:10:13. This time has since been bettered by six North Americans. Stefan Fekner of Canada beat it by the amazingly close time of just 3 seconds, and Don Paul, who had dropped out early in the 1981 Chicago race, beat that by a hair's breadth of just 2 seconds in a later race!

My times for both the 50-mile and 100-kilometer were then, and still are now, masters (over age forty) world records.

While writing this book, I contacted Andy Milroy (former technical director of the International Association of Ultrarunners), who e-mailed me the following:

> To put your 6:38:21 100 km time into worldwide context . . . Because of the uncertainty surrounding the 6:36:57 (then a possible world road best) by Richard Chouinard of Canada at the Montmagny race on 21/22 July, 1979, your 6:38:21 at Chicago on 4 Oct. 81 was the best certified road 100 km mark in the world—effectively the world record. Based on the documentation that was available on the Montmagny mark as compared with the Chicago mark, the latter would have had priority.

I hasten to add another point of worldwide context, namely, there are always better performances. My race time, for example, did not come close to the sparkling brilliance of Scotland's Don Ritchie, who on October 28, 1978, ran a track 100-kilometer in 6:10:20, at the Crystal Palace in London, and who in his younger years, set 15 world records. After turning forty, he ran from Turin to St. Vincent, Italy, in 6:36:02 (point-to-point races may count as "notable performances," but they are not eligible for world records). His many stellar performances," cancel all ambitions I might have had for world bragging rights.

In retrospect, the race is a metaphor for my life. Our lives are influenced by our evolutionary past, by experience, and by our minds. There are times when we take life as it comes, and there are other times when we apply maximum effort to try to achieve a certain result. Throughout, it is all an adventure that we like to look back on with pride. We don't really *know* if what we're doing will get us where we want to go. I didn't really *know* whether what I was doing to prepare for the race was right. As when choosing the ideal mate, a course of study, or a training regimen, we take calculated risks. In retrospect, I made foolish mistakes. I had not, for example, taken the findings from the birds into account; I should have rested up more, carbo depleted less, and taken a little protein on the run. Undoubtedly, I made many more mistakes that I still don't even realize. Despite my likely errors, I was asked to record my experiment of one in two ultrarunning training manuals, as were other ultrarunners. One learns more from one's mistakes than from one's successes, and I therefore concentrate on them.

I drank one and a half gallons of cranberry juice during the race but still lost 8 pounds of body weight, and since my kidneys had shut down—I never once urinated—I must have lost about 20 pounds of fluid through sweat. Feeling that cranberry juice might have been a magic elixir, I used it in a 50-mile race that was held on a chilly late fall day in Maine. I did not sweat as much and urinated instead. It cost me much time. I learned that the juice is highly diuretic. For another race, I trained even harder—peaking at 200 miles per week—for a 24-hour race in North Carolina. I again used cranberry juice. This time it tasted so foul when I tried to drink it on the run that I could not get it down. I was reduced to drinking water, and the water contaminated by the mere taste of the juice remaining in my squeeze bottle made it soon impossible for me to stomach even that. I hit the wall at 32 miles and dropped out. I learned that I should have read the fine print on the label. This cranberry juice, it turned out later, lacked corn syrup additive and had an artificial sweetener instead; the lack of endurance came from the lack of racing fuel.

The taste aversion could have been from having trained

my body during the race that the pain from running comes from the juice being ingested. That is, I would have been nauseated with *any* cranberry juice, because my body remembered from the race that the more cranberry juice I drank, the more I felt pain, and so my body then tagged the pain as coming from cranberry rather than running. In the same way rats, once made to feel pain or ill by food with poison, develop a taste aversion to the flavor of the food that made them ill (for example, when receiving radiation while eating that food). I had, apparently, *unconsciously* remembered the pain experienced during the race, although I no longer had a clear *sensory* memory of it. Our inability to retain a *sensory* memory of pain may be a psychological adaptation allowing us to repeat the hunt, to not become immobilized, and to repeat the race again and again. (I went on to set American records at the 100 mile, and most miles run in 24 hours before it sunk in.) In contrast, I remember still the exquisite rapture from childhood of capturing my first tiger beetle, holding a baby bird, and a thousand other pleasures that motivate and empower me to act even now.

I was invited to run in the Greek Spartathlon, a 152-mile race from Athens to Sparta. This time I felt pressure to win, and probably due to my previous success and short memory, I took off with the European antelopes on that still dark dawn in Athens. When I got into the steep mountains before Sparta, I stopped to walk. That felt like quitting. I had not yet realized that walking breaks can be part of ultrarunning strategy. This time my friend Krolewicz, the camel, who was also in the American contingent, beat me handily; he finished. I had not heeded the example of the camels and frogs, which pace themselves and do it right. Nor had I invoked caution from the example of the deer, which do it wrong by sprinting when they are chased by an endurance predator.

Success is different things to different people, as I was reminded on a recent trip across the country. Our family stopped at Red Rock State Park near Gallup, New Mexico. An Indian man jogged by us up the canyon while we were picnicking. When he came back down, I intercepted him for a chat. We discovered we were both distance runners. He had recently completed six marathons even though at one time he weighed 250 pounds, smoked heavily, and had high blood pressure and a drinking problem. Running had saved his life. "Every time I cross the finish line," he told me, "I win." He had won big. His brain had done it. He thought he could do it, and did it. He could have been dead wrong. I also feel that I owe much to my running—my education, my health, and maybe my life.

In Chicago I'd done the best that I knew how at the time. That's what mattered to me. As I said, when I finished the race I didn't even know what record, if any, I might have set. I didn't consciously dwell on the race until nineteen years later, when I watched my son in a high school cross-country meet. It sent shivers down my spine. Memories kept coming and it all seemed as if it had happened yesterday. And so I began to write, lest I lose the experience that had been so costly and so precious to me. I wanted to pass it on, and to retrieve the running experience anew I also started to relive it; I began a training regimen to try for some age-group (over sixty) records.

When Bushmen kill an eland, wildebeest, or some other antelope, they share the meat with their friends. They gather around the glowing embers of their campfire and talk deep into the night about the hunt. When they are not hunting, they are talking about hunting, reliving their experience.

I believe our common hunter's heart is the ability to

impart value far in excess of what seems practical. That's dreaming. That's a large part of what makes us human. If modern runners were drawn around a campfire in a warm African night, they would, like any Bushmen, poke the embers and relive the run all the way to the finish line and beyond. That's what I've tried to do here.

References

Chapter 1

Urquhart, F. A. 1987. *Monarch Butterfly: International Traveler.* Chicago: Nelson-Hall.

Chapter 2

Martin, D. E., and R. W. H. Gynn, 1979. *The Marathon Footrace.* Springfield, Ill.: Charles C. Thomas.

Chapter 3

Costill, D. L. 1979. *A Scientific Approach to Distance Running.* Track & Field News. (No town given.)

Chapter 4

Bartholomew, G. A., and B. Heinrich. 1978. Endothermy in African dung beetles during flight, ball making, and ball rolling. *J. Exp. Biol.* 73:65–83.

Heinrich, B., and G. A. Bartholomew. 1979. Roles of endothermy and size in inter- and intraspecific competition for elephant dung in an African dung beetle, *Scarabaeus laevistriatus. Physiol. Zool.* 52: 484–96.

Morgan, K. R. 1985. Body temperature regulation and terrestrial activity in the ecothermic beetle *Cicindela tranquebarica. Physiol. Zool.* 58:29–37.

Chapter 5

Hadley, M. E. 1996. *Endocrinology*. Upper Saddle River, N.J.: Prentice Hall.

Hylan, D. A. 1990. *Physiology of Sport*. New York: Paragon.

Nijhout, H. F. 1994. *Insect Hormones*. Princeton: Princeton University Press.

Speck, F. G. 1940. *Penobscot Man: The Life History of a Forest Tribe in Maine.* London: Oxford University Press.

Tauber, M. J., C. A. Tauber, and S. Masaki. 1986. *Seasonal Adaptation of Insects*. New York: Oxford University Press.

Chapter 6

Cook, J. R., and B. Heinrich. 1965. Glucose vs. acetate metabolism in *Euglena. J. Protozool.* 12:581–84.

———. 1968. Unbalanced respiratory growth of *Euglena. J. Gen. Microbiol.* 53:237–51.

Costill, D. L. 1970. Metabolic responses during distance running. *J. Applied Physiol.* 28:251–53.

Heinrich, B., and J. R. Cook. 1967. Studies on the respiratory physiology of *Euglena gracilis* cultured on acetate or glucose. *J. Protozool.* 14:548–53.

Noakes, T. 1985. *The Lore of Running*. Cape Town: Oxford University Press.

Schmidt-Nielson, K. 1990. *Animal Physiology: Adaptation and Environment*. Cambridge: Cambridge University Press.

Wilmore, J. H. 1982. *Training for Sport and Activity: The Physiological Basis of the Conditioning Process*. Boston: Allyn and Bacon.

Chapter 7

Bramble, D. M., and D. R. Carrier. 1983. Running and breathing in mammals. *Science* 219:251–56.

Fixx, J. F. 1977. *The Complete Book of Running.* New York: Random House.

Heinrich, B. 1970. Nervous control of the heart during thoracic temperature regulation in a sphinx moth. *Science* 169:606–7.

———. 1970. Thoracic temperature stabilization in a free-flying moth. *Science* 168:580–83.

———. 1971. Temperature regulation of the sphinx moth, *Manduca sexta*. *J. Exp. Biol.* 54:141–66.

———. 1976. Heat exchange in relation to blood flow between thorax and abdomen in bumblebee. *J. Exp. Biol.* 64:561–85.

———. 1979. Keeping a cool head in honeybee thermoregulation. *Science* 205:1269–71.

Chapter 8

Able, K. P., ed. 1999. *Gatherings of Angels*. Ithaca: Comstock Books.

Baird, J. 1999. Returning to the tropics: The epic autumn flight of the blackpoll warbler. In *Gatherings of Angels*. K. P. Able, ed. Ithaca: Comstock Books.

Berthold, P. 1996. *Control of Bird Migration*. London: Chapman and Hall.

Ens, B. J., T. Piersma, W. J. Wolff, and L. Zwarts, eds. 1990. Homeward bound: Problems waders face when migrating from the Bane d'Arguin, Mauritania, to their northern breeding grounds in spring. *Ardea* 78:1–363.

Gonzalez, N. C., and R. M. Fedde. 1988. *Oxygen Transport from Atmosphere to Tissues.* New York: Plenum.

Harrington, B. A. 1999. The hemispheric globetrotting of the white-rumped sandpiper. In *Gatherings of Angels.* K. P. Able, ed. Ithaca: Comstock Books.

Piersoma, T., and N. Davidson. 1992. *The Migration of Knots.* Wader Study Group Bulletin 64. Petersborough, U.K.: Monkstone House.

Piersma, T., A. Koolhaas, and A. Dekinga. 1993. Interactions between stomach structure and diet choice in shorebirds. *Auk* 110:552–64.

Schmidt-Nielsen, K. 1992. *How Animals Work*. London: Cambridge University Press.

Tucker, V. A. 1968. Respiratory exchange and evaporative water loss in the flying budgerigar. *J. Exp. Biol.* 48:67–87.

Chapter 9

Burney, D. A. 1993. Recent animal extinctions: Recipes for disaster. *Am. Sci.* 81:530–41.

Byers, J. A. 1984. Play in ungulates. In *Play in Animals and Humans,* ed. P. K. Smith. Oxford: Basil Blackwell.

———. 1997. *American Pronghorn: Social Adaptations and the Ghosts of Predators Past.* Chicago and London: University of Chicago Press.

Eyestone, E. 2000. "Man vs. Horse." *Runner's World* (November).

Kurtén, B., and E. Anderson. 1980. *Pleistocene Mammals of North America.* New York: Columbia University Press.

Lindstedt, S. L., J. F. Hokanson, D. J. Wells, S. D. Swain, H. Hoppeler, and V. Navarro. 1991. Running energetics in the pronghorn antelope. *Nature* 353:748–49.

Mech, L. D. 1970. *The Wolf: The Ecology and Behavior of an Endangered Species.* New York: Doubleday.

Nobokov, P. 1981. *Indian Running: Native American History and Tradition.* Santa Fe, N.M.: Aneburt City Press.

Price, E. O. 1984. Behavioral aspects of animal domestication. *Quarterly Rev. Biol.* 59:1–32.

Stuart, A. J. 1991. Mammalian extinctions in the late Pleistocene of northern Eurasia and North America. *Biol. Rev.* 66:453–62.

Turbak, G. 1995. *Pronghorn: Portrait of the American Antelope.* Flagstaff, Ariz.: Northland Publishing Co.

Webb, S. D. 1977. A history of savanna vertebrates in the New World. Part I, North America. *Annu. Rev. Ecol. & Syst.* 8:355–80.

Chapter 10

Dagg, A. J. 1974. The locomotion of the camel (*Camelus dromedarius*). *J. Zool.* 174:67–68.

Denis, F. 1970. Observations sur le compartement du dromadaire. Thesis. Faculté des Sciences de l'Université de Nancy.

Gauthier-Pilters, H., and A. I. Dagg. 1981. *The Camel: Its Evolution, Ecology, Behavior, and Relationship to Man.* Chicago: University of Chicago Press.

Louw, G. 1993. *Physiological Animal Ecology.* Essex, U.K.: Longman Scientific and Technical.

McKnight, T. L. 1969. *The Camel in Australia.* Carlton: Melbourne University Press.

Perk, R. F. 1963. The camel's erythrocyte. *Nature* 200: 272–73.

———. 1966. Osmotic hemolysis of the camel's erythrocytes. *J. Exp. Zool.* 163:241–46.

Schmidt-Nielsen, K. 1959. The physiology of the camel. *Sci. Am.* 201:140–51.

———. 1964. *Desert Animals: Physiological Problems of Heat and Water.* Oxford: Clarendon Press.

Schmidt-Nielsen, K., E. C. Crawford, A. E. Newsholme, K. S. Rawson, and H. T. Hammel. 1967. Metabolic rate of camels: Effect of body temperature and dehydration. *Amer. J. Physiol.* 212:341–46.

Schmidt-Nielsen, K., B. Schmidt-Nielsen, T. R. Houpt, and S. A. Jarnum. 1956. The question of water storage in the stomach of the camel. *Mammalia* 20:11–15.

———. 1956. Water balance of the camel. *Amer. J. Physiol.* 185:185–94.

Chapter 11

Billings, D. 1984. Aerobic efficiency in ultrarunners. *Ultrarunning,* November 1984, 24–25.

Davies, C. T. M. 1981. Physiology of ultra-long distance running. *Medicine and Sport* 13:53–63.

Taigen, T. L., and K. D. Wells. 1985. Energetics of vocalization by an anuran amphibian (*Hyla versicolor*). *J. Comp. Physiol.* 155:163–70.

Taigen, T. L., K. D. Wells, and R. L. Marsh. 1985. The enzymatic basis of high metabolic rates in calling frogs. *Physiol. Zool.* 58:719–26.

Wells, K. D., and T. L. Taigen. 1986. The effect of social interactions on calling energetics in the gray treefrog (*Hyla versicolor*). *J. Behav., Ecology, and Sociobiology* 19:9–18.

Chapter 12

Alexander, R. M. 1984. Elastic energy stores in running vertebrates. *Amer. Zool.* 24:85–94.

———. 1988. *Elastic Mechanisms in Animal Movement.* Cambridge, U.K.: Cambridge University Press.

Darwin, C. R. 1859. *On the Origin of Species by Means of Natural Selection, or The Preservation of Favored Races in the Struggle for Life.* London: John Murray.

Gordon, D. G. 1996. *The Compleat Cockroach: A Comprehensive Guide to the Most Despised (and Least Understood) Creatures on Earth.* Berkeley, Calif.: Ten Speed Press.

Ker, R. F., M. B. Bennett, S. R. Bibby, R. C. Kester, and R. McN. Alexander. 1987. The spring in the arch of the human foot. *Nature* 325:147–49.

McMahon, T. A. 1987. The spring in the human foot. *Nature* 325:108–9.

McMahon, T. A., and P. R. Greene. 1979. The influence of track compliance on running. *J. Biomech.* 12:893–904.

Vogel, S. 1998. *Cat's Paws and Catapults.* New York: W. W. Norton.

Chapter 13

Andrade, M. C. B. 1996. Sexual selection for male sacrifice in the Australian redback spider. *Science* 271:70–72.

Bennett, W. C., and R. M. Zingg. 1935. *The Tarahumara: An Indian Tribe of Northern Mexico.* Chicago: University of Chicago Press.

Borta, W. M. 1985. Physical exercise as an evolutionary force. *J. Human Evolution.* 14:145–55.

Bramble, D. M., and D. R. Carrier. 1983. Running and breathing in mammals. *Science* 219: 251–56.

Burney, D. A. 1993. Recent animal extinction: Recipes for disaster. *American Scientist* 81:530–41.

Caputa, M. 1981. Selective brain cooling: An important component of thermal physiology. *Contributions to Thermal Physiology* 32:183–92.

Carrier, D. R. 1984. The energetic paradox of human running and hominid evolution. *Current Anthropology* 24 (4):483–95.

Dawson, T., J. D. Robertshaw, and C. R. Taylor. 1974. Sweating in the kangaroo: A cooling mechanism during exercise, but not in the heat. *Amer. J. Physiol.* 227:494–98.

Falk, D. 1990. Brain evolution in homo: The "radiator" theory. *Behavioral and Brain Sciences* 13:333–86.

Gaesser, C. A., and G. A. Brooks. 1980. Glycogen depletion following continuous and intermittent exercise to exhaustion. *Journal of Applied Physiology* 49:727–28.

Hawkes, K. 1991. Showing off: Tests of an hypothesis about men's foraging goals. *Ethology and Sociobiology* 12:29–54.

Heinrich, B. 1970. Thoracic temperature stabilization by blood circulation in a free-flying moth. *Science* 168:580–82.

———. 1972. Thoracic butterflies in the field near the equator. *Comp. Biochem. Physiol.* 43A:459–67.

———. 1993. *The Hot-Blooded Insects.* Cambridge: Harvard University Press.

———. 1996. *The Thermal Warriors.* Cambridge: Harvard University Press.

Johanson, D., and M. Edey. *Lucy: The Beginnings of Humankind.* New York: Simon & Schuster.

Kaplan, H., and K. Hill. 1985. Hunting ability and reproduction success among male Aché foragers: Preliminary results. *Current Anthropology* 26:131–33.

Kessel, E. L. 1955. The mating activities of balloon flies. *Syst. Zool.* 4:97–104.

Lee, R. B., and I. DeVore, eds. 1968. *Man the Hunter.* Chicago: Aldine.

Lee, R. B. 1979. *The !Kung San: Men, Women, and Work in a Foraging Society.* Cambridge: Cambridge University Press.

Leonard, W. R., and M. L. Robertson. 2000. Ecological correlates of home range variation in primates: Implications for hominid evolution. In *On the Move: How Animals Travel in Groups,*

S. Boinski and P. A. Garber, eds. Chicago and London: University of Chicago Press.

Louw, G. 1993. *Physiological Animal Ecology.* Essex, U.K.: Longman Scientific and Technical.

Lowie, R. H. 1924. Notes on Shoshonean ethnography. *Anthropological Papers of the American Museum of Natural History* 20, part 3.

May, M. 1976. Thermoregulation and adaptation to temperature in dragonflies (*Odonata: Anisoptera*). *Evol. Monogr.* 46:1–32.

McCarthy, F. D. 1957. *Australian Aborigines: Their Life and Culture.* Melbourne: Colorgravure Publications.

Newman, R. W. 1970. Why man is such a sweaty and thirsty naked animal: A speculative review. *Human Biology* 42:12–27.

Pennington, C. W. 1963. *The Tarahumara of Mexico.* Salt Lake City: University of Utah Press.

Poulten, E. B. 1913. Empidae and their prey in relation to courtship. *Entomol. Mo. Mag.* 49:177–80.

Rudman, P. S., and H. M. McHenry. 1980. Bioenergetics and the origin of hominid bipedalism. *American Journal of Physical Anthropology* 52:103–6.

Schaller, G. B., and G. R. Lowther. 1969. The relevance of carnivore behavior to the study of early hominids. *Southwestern Journal of Anthropology* 25:307–41.

Schapera, I. 1930. *The Khoisan People of South Africa: Bushman and Hottentots.* London: Routledge and Kegan Paul.

Shoemaker, V. H., K. A. Nagy, and W. R. Costa. 1976. Energy utilization and temperature regulation by jackrabbits (*Lepus californicus*) in the Mojave Desert. *Physiol. Zool.* 49:364–75.

Sollas, W. J. 1924. *Ancient Hunters and Their Modern Representatives.* New York: MacMillan.

Stanford, C. B. 1995. To catch a colobus. *Natural History* 1:48–54.

———. 1999. *The Hunting Apes: Meat Eating and the Origins of Human Behavior.* Princeton: Princeton University Press.

Steudel, K. 1996. Limb morphology, bipedal gait, and the energetics of hominid locomotion. *American Journal of Physical Anthropology* 99:345–55.

Strum, S. C. 1981. Processes and products of change: Baboon predatory behavior at Gilgil, Kenya. In *Omnivorous Primates: Gathering and Hunting in Human Evolution,* ed. R. S. O. Harding and G. Teleki. New York: Columbia University Press.

Taylor, C. R., N. C. Heglund, and G. M. O. Maloiy. 1982. Energetics and mechanisms of terrestrial locomotion. *J. Exp. Biol.* 97:1–21.

Taylor, C. R., and V. J. Rowntree. 1973. Running on two or four legs: Which consumes more energy? *Science* 179:186–87.

———. 1973. Temperature regulation and heat balance in running cheetahs: A strategy for sprinters? *Amer. J. Physiol.* 224:848–51.

Toolson, E. C. 1987. Water profligacy as an adaptation to hot deserts: Water loss rates and evaporation cooling the Sonoran Desert cicada, *Diceroprocta apache* (Homoptera: Cicadidae). *Physiol. Zool.* 60:379–85.

Wannenburgh, A. 1979. *The Bushmen.* Cape Town: C. Struik.

Washburn, S. L., and C. Lancaster. 1968. The evolution of hunting. In *Man the Hunter,* ed. R. B. Lee and I. DeVore. Chicago: Aldine.

Wheeler, P. R. 1984. The evolution of bipedalability and loss of functional body hair in hominids. *Journal of Human Evolution* 13:91–98.

———. 1991. Thermoregulatory advantages of hominid bipedalism in open equatorial environments: The contribution of increased heat loss and cutaneous evaporative cooling. *Journal of Human Evolution.* 21:107–15.

Wolpoff, M. H. 1980. *Paleoanthropology.* New York: Knopf.

Wrangham, R. W., J. H. Jones, G. Laden, D. Pilbeam, and N. Conklin-Brittain. 1999. The raw and the stolen: Cooling and the ecology of human origins. *Current Anthropology* 40:567–94.

Chapter 14

Adler, N. T. 1981. *Neuroendocrinology of Reproduction: Physiology and Behavior*. New York: Plenum Press.

Bale, J., and J. Sang. 1996. *Kenyan Running.* London: Frank Cass.

Beck, S. D. 1980. *Insect Photoperiodism*. New York: Academic Press.

Berg-Schosser, D. 1984. *Tradition and Change in Kenya: Comparative Analysis of Seven Major Ethnic Groups.* Paderborn: Ferinand Schöningh.

Cobb, W. M. 1936. Race and runners. *Journal of Health and Physical Education* 1:3–7, 53–55.

Daniels, J. 1975. Science on the altitude factor. In *The African Running Revolution,* D. Prokop, ed. Mountain View, Calif.: World Publications.

Derderian, T. 1994. *Boston Marathon: The History of the World's Premier Running Event.* Champaign, Ill.: Human Kinetics.

Derr, M. 1996. The making of a marathon mutt. *Natural History* 3:35–40.

Donovan, C. M., and G. A. Brooks. 1983. Endurance training affects lactate clearance, not lactate production. *Amer. J. Physiol.* 244:E82–E92.

Hoberman, J. 1952. *Mortal Engines: The Science of Performance and the Dehumanization of Sport.* New York: Free Press.

Saltin, B., et. al. 1995. Aerobic exercise capacity at sea level and at altitude in Kenyan boys, junior and senior runners compared with Scandinavian runners. Scand. Journ. *on Science in Sports* 5(4):209–21.

Wiggin, D. 1999. "Great speed but little stamina": The historical debate over black superiority. *Journal of Sports History* 16(2):158–85.

Wehner, R. A., A. C. Marsh, and S. Wehner. 1992. Desert ants on a thermal tightrope. *Nature* 357:586–87.

Chapter 15

Revkin, A. C. 1989. Sleeping beauties: The bear's strategies of getting through the winter. *Discover* (April): 62–65.

Chapter 16

Allport, S. 1999. *The Primal Feast.* New York: Harmony Books.

Battley, P. F., T. Piersma, M. W. Dietz, S. Tang, A. Dekinga, and K. Hulsman. 1999. Empirical evidence for differential organ reductions during trans-oceanic bird flight. *Proc. Royal Soc. London* B 267:191–95.

Biebach, H. 1998. Phenotypic organ flexibility in garden warbler *Sylvia borin* during long-distance migration. *J. Avian Biol.* 29: 529–35.

Karasov, W. H., and B. Pinshow. 1998. Changes in lean mass and in organs of nutrient assimilation in long-distance passerine migrant at a spring-time stopover site. *Physiol. Zool.* 71:435–48.

Larsen, C. S. 1997. *Bioarcheology: Interpreting Behavior from the Human Skeleton.* Cambridge: Cambridge University Press.

Chapter 17

Heinrich, B. 1979. *Bumblebee Economics*. Cambridge: Harvard University Press.

Piersma, T., G. A. Gudmundsson, and K. Lilliendahl. 1999. Rapid changes in size of different functional organ and muscle groups during refueling in a long-distance migrating shorebird. *Physiol. Biochem. Zool.* 72:405–16.

Chapter 18

Battley, P. F., T. Piersma, M. W. Dietz, S. Tang, A. Dekinga, and K. Hulsman. 1999. Empirical evidence for differential organ reductions during trans-oceanic bird flight. *Proc. Royal Soc. London* B 267:191–95.

Gaesser, C. A., and G. A. Brooks. 1980. Glycogen depletion following continuous and intermittent exercise to exhaustion. *Journal of Applied Physiology* 49:727–28.

Hochachka, P. W., and G. N. Somero. 1984. *Biochemical Adaptation.* Princeton: Princeton University Press.

Hollaszy, J. O., and F. W. Booth. 1976. Biochemical adaptations to endurance in muscle. *Ann. Rev. Physiol.* 38:273–91.

Jacobs, B. L., H. van Praag, and F. H. Gaze. 2000. Depression and the birth and death of brain cells. *American Scientist* 88:340–54.

McArdle, W. D., F. I. Katch, and V. L. Katch. 1991. *Exercise Physiology, Energy, Nutrition, and Human Performance.* 3rd ed. Philadelphia and London: Lea and Febiger.

Noakes, T. 1985. *The Lore of Running.* Cape Town: Oxford University Press.

Piersma, T., L. Bruinzeel, R. Drent, M. Kersten, J. V. der Meer, and P. Wiersma. 1996. Variability in basal metabolic rate of a long-distance migrant shorebird (red knot, *Calidris canutus*) reflects shifts in organ sizes. *Physiol. Zool.* 69:191–217.

Sleamaker, R. 1989. *Serious Training for Serious Athletes*. Champaign, Ill.: Leisure Press.

Chapter 19

Steinberg, D., and J. C. Khoo. 1977. Hormone-sensitive lipase of adipose tissue. *Fed. Proc.* 36:1986–90.

Index